Frontiers and Progress in Multi

Series editors

Mewes Dieter, Hannover, Germany
Lixin Cheng, Aarhus C, Denmark

For further volumes:
http://www.springer.com/series/11688

Lixin Cheng
Editor

Frontiers and Progress in Multiphase Flow I

Springer

Editor
Lixin Cheng
Faculty of Science and Technology
Department of Engineering
Aarhus University
Aarhus C
Denmark

ISBN 978-3-319-35526-9 ISBN 978-3-319-04358-6 (eBook)
DOI 10.1007/978-3-319-04358-6
Springer Cham Heidelberg New York Dordrecht London

Printed on acid-free paper

Springer is part of Springer Science+Business Media (www.springer.com)

Contents

Chapter 1
Selected Aspects of Thermal-Hydraulics Modelling in Two-Phase Flows with Phase Change in Minichannels

Dariusz Mikielewicz

Abstract In the chapter some issues of thermal-hydraulics modeling of two-phase flows in minichannels with change of phase are presented. These encompass the common modeling of flow boiling and flow condensation using the same expression. Approaches to model these two respective cases show, however, that experimental data show different results to those obtained by methods of calculation of heat transfer coefficient for respective cases. Partially that can be devoted to the fact that there are non-adiabatic effects present in both types of phase change phenomena which modify the pressure drop due to friction, responsible for appropriate modelling. The modification of interface shear stresses between flow boiling and flow condensation in case of annular flow structure may be considered through incorporation of the so called blowing parameter, which differentiates between these two modes of heat transfer. On the other hand, in case of bubbly flows, the generation of bubbles also modifies the friction pressure drop by the influence of heat flux. Presented are also the results of a peculiar M-shape distribution of heat transfer coefficient specific to flow boiling in minichannels. Finally, some attention is devoted to mathematical modeling of dryout phenomena. A five equation model enabling determination of the dryout location is presented, where the mass balance equations for liquid film, droplets and gas are supplemented by momentum equations for liquid film and two-phase core.

Keywords Flow boiling · Flow condensation · Pressure drop · Non-adiabatic effects · Minichannels · Dryout · M-shape distribution of heat transfer coefficient

D. Mikielewicz (✉)
Faculty of Mechanical Engineering, Department of Energy and Industrial Apparatus, Gdansk University of Technology, ul. Narutowicza 11/12, 80-233 Gdansk, Poland
e-mail: Dariusz.Mikielewicz@pg.gda.pl

L. Cheng (ed.), *Frontiers and Progress in Multiphase Flow I*,
Frontiers and Progress in Multiphase Flow, DOI: 10.1007/978-3-319-04358-6_1,

Nomenclature

B	Blowing parameter, $B = 2\vartheta_0/(c_f u_\infty)$
Bo	Boiling number, $Bo = q/(Gh_{LG})$
c_f	Friction factor
c_p	Specific heat, J/kg K
C	Concentration
Con	Constraint number, $Con = (\sigma/g/(\rho_l - \rho_v))^{0.5}/d$
d	Channel inner diameter, m
E	Energy dissipation, W/m^3
f	Friction factor
f_1, f_{1z}	Functions
F	Enhancement factor, function
g	Gravity, m/s^2
G	Mass flux, kg/m^2 s
h	Enthalpy, J/kg
h_{lv}	Latent heat of evaporation, J/kg
k	Mass transfer coefficient, m/s^2
l	Bubble characteristic length, m
L	Channel length, m
p	Pressure, Pa
P	Correction in Eq. (1.46)
Pr	Prandtl number
q	Heat flux, W/m^2
R	Two-phase flow multiplier
R_{MS}	Two-phase flow multiplier due to Müller-Steinhagen and Heck [17]
Re	Reynolds number
S	Suppression factor
x	Quality
X_{tt}	Martinelli parameter
T	Temperature, K
u, w, ϑ_0	Velocity, m/s
z	Distance along the channel, m
α	Heat transfer coefficient, W/m^2 K
δ	Thickness of liquid film, m
ξ	Drag coefficient
λ	Thermal conductivity, W/m K
μ	Dynamic viscosity, Pa s
ρ	Density, kg/m^3
σ	Surface tension, N/m
τ	Shear stress, Pa

Subscripts

c	Core
cb	Convective boiling
D	Deposition
E	Entrainment
f	Forced flow, liquid film
G	Saturated vapour
i	Interface
k	Droplets
kr	Critical
L	Liquid
LO	Liquid only
mt	Mass transfer
PB	Pool boiling
TP	Two-phase flow
TPB	Two-phase boiling
TPK	Two-phase condensation
0	Beginning of annular flow
∞	Undisturbed flow

1.1 Introduction

Two-phase flows with change of phase, namely evaporation and condensation are amongst most important industrial processes which are used primarily in power engineering, chemical technology as well as other areas of contemporary technology and they have huge practical significance, just to mention perspective applications in removal of large heat fluxes in electronics. Conversion of one phase into the second one may proceed equally on the wall and inside the phase, however usually the first of these is found in technical applications. These phenomena are more complicated than issues of convection in case of single phase flows. Their mathematical description requires thus knowledge of the flow structure of both phases together with interfacial phenomena. Hence, existing theories provide only simplified description of the issues related to these phenomena and require experimental validation.

Flow boiling and flow condensation are often regarded as two opposite or symmetrical phenomena involving the change of phase. There is a temptation to describe both these phenomena with one only correlation, however such method has yet to be put forward. From amongst the structures present in flow boiling and flow condensation at least the annular flow structure seems to be suitable to the common modeling. However, the shear stress acting between vapour phase and liquid phase is not the same in the respective cases, i.e. flow boiling and flow condensation. Most of modeling of heat transfer in case of condensation inside channels relates the heat transfer coefficient to the friction coefficient. Such approach is rather not commonly used in case of flow boiling. In the latter case, all existing approaches are either the empirical fits to experimental data or form an attempt to combine two major influences to heat transfer, namely the convective flow boiling without bubble generation and nucleate boiling. Generally that is done in a linear or non-linear manner. Alternatively, there is a group of modern approaches based on models which start from modeling a specific flow structure and in such way postulate more accurate flow boiling models, usually pertinent to annular or bubbly flows.

Generally, the non-adiabatic effects modify the friction pressure drop term and subsequently the heat transfer coefficient. That is a major reason why up to date approaches to common modelling of flow boiling and flow condensation as symmetric are failing in that respect. In author's opinion the way to solve that problem is to incorporate appropriate mechanisms into the friction pressure drop term responsible for modification of shear stresses at the vapour-liquid interface, different for annular flow structure and different for the other ones. Postulated here suggestion of considering the so called "blowing parameter" in annular flow explains partially the mechanism of liquid film thickening in case of flow condensation and thinning in case of flow boiling. In other flow structures, for example the bubbly flow, there can also be identified other effects, which have yet to attract sufficient attention in literature. One of such effects is the fact that the two-phase pressure drop is modelled in the way that the influence of applied heat flux is not considered. The effect of heat flux is especially important in case of small qualities where the non-adiabatic effect is pronounced, which will be shown in later.

Considerations presented in the paper relate both to the case of flow boiling and flow condensation in conventional channels as well as small diameter ones. The transition between conventional channels and small channels was suggested by Kandlikar [1]. He proposed that channel diameters smaller than 3 mm should be regarded as minichannels and also devised a limit for microchannels to be smaller than 600 μm. There are also criteria based on more physical grounds just to mention the one based on the value of the Constraint number *Con*, Kew and Cornwell [2] ($Con = (\sigma/(g(\rho_l - \rho_v)D_h^2)^{0.5})$. If a value of that number in the flow is greater than 0.5 then the two-phase flow regarded as then one in a minichannel. Such a distinction between macro- and micro-scale by the threshold diameter of 3 mm is adopted due to the lack of a well-established theory as pointed out by Cheng et al. [3, 4]. Author devoted all the possible attention that the modeling presented is applicable to the whole range of quality variation in both cases of condensation and boiling. One can expect that bubble nucleation in flow boiling renders the process of heat transfer not to have its counterpart in the condensation inside tubes. Similarly in as the case of inside tube condensation, where the collapse of bubbles to form a continuous liquid structure is the condensation specific phenomenon. The form of two-phase flow multiplier, which is a major factor in modeling presented here, should be capable of capturing all the mentioned above cases, however in case of some fluids more studies will be required to devise an appropriate version of the two-phase flow multiplier. Situation seems to be a little less complex in the case of flow boiling in minichannels and microchannels. In such flows the annular flow structure is dominant for most qualities, Thome and Consolini [5], as the bubbly/slug flows are found for vapour qualities not exceeding $x = 0.1$. In case of the annular flow the heat transfer coefficient is primarily dependent on the convective mechanism.

The objective of the present book is to present some of the advances in modeling of flow boiling and condensation. First, the capability of the flow boiling model, developed earlier by the author in [6] to model also the condensation inside tubes [7] is discussed. As mentioned above, some modifications to that model, developed recently, will also be presented. That regards the "blowing parameter", which modifies the shear stress in annular flow and the influence of heat flux in bubbly flow. The predictions obtained using authors own model will be confronted with obtained in-house experimental data collected recently. Secondly, some discussion regarding the M-shape distribution of heat transfer coefficient in flow boiling is presented on the basis of in-house experimental data.

Another issue considered in the chapter is the phenomenon of dryout, which is the condition representing the breaking of continuous liquid contact with the heated surface in a flow channel at moderate to high qualities. It follows the gradual decrease of liquid fraction due to evaporation or boiling of the liquid film. The flow pattern before the dryout condition is usually annular with the vapor phase occupying most of the core with the liquid flowing as a thin film along the channel wall. To simulate the dryout location the original five-equation model was developed. It is based on the liquid balance in the film and in the core and also the vapor balance in the annular two-phase flow. These equations are supplemented by equations of momentum balance in liquid and vapour core. The model is dedicated to the dryout phenomena modeling in both the conventional and mini-channels.

1.2 Overview of Flow Structures and Basic Parameters Defining Two-Phase Flow

The objective of studying two-phase flow is to determine the heat transfer and pressure drop characteristics of a given flow. The adiabatic two-phase flow is different from the diabatic one, i.e. the one with heat addition. Modelling of adiabatic two-phase flow presents a serious challenge, whereas the flow with heat transfer is even a more complex issue. Application of heat flux causes development of boiling and the flow development through various flow patterns. Flow pattern affects the hydrodynamics, such as the pressure drop along the flow path which further affects the heat transfer characteristics.

In case of flow boiling the wall must be first superheated to a value that depends on its microgeometry and wettability in order to activate nucleation sites, similarly as in pool boiling. In subcooled region generated vapor bubbles condense almost immediately in the bulk of the flow. The subcooled boiling zone extends to the location where temperature across the channel reaches saturation temperature.

Following the bubbly flow regime the fraction of the channel cross section occupied by vapor becomes fairly large and annular flow structure is obtained with the liquid film flowing on the channel wall and vapor core being in the centre. Within this regime heat transfer occurs directly from the wall to the interface and vaporization process on interface occurs. Nucleate boiling is then suppressed. As the vapor-liquid mixture flows to higher vapor quality region the quantity of liquid in the flow decreases and wall dryout sets in, which is accompanied by the sudden rise of wall temperature. Transition occurs to the dispersed or mist-type flow of the mixture. Presented description of the change of regimes with the increase of vapor quality x is presented in a simplified manner in Fig. 1.1.

In horizontal channels, due to gravity, a wider variety of regimes exist than in vertical channels. Determination of the most probable flow regime can be done using the so-called regime charts or flow maps. The most commonly used diagrams for conventional size channels are due to Baker [8] and Taitel-Dukler [9] for horizontal flows and the Hewitt-Roberts [10] and Oshinowo-Charles [11] for upward and downward vertical flows respectively. Two-phase flows in minichannels under conditions where inertia is significant have been experimentally investigated rather extensively. Experiments generally show that, with the exception of stratified flow, which does not occur when $D_h < 1$ mm with air/water-like fluid pairs, all other major flow regimes (bubbly, slug, churn, annular, etc.) can occur in minichannels. The flow regimes and their parameter ranges are also similar for vertical and horizontal channels, and they are insensitive to channel orientation. Examples of such maps are due to Triplett et al. [12] and Chung and Kawaji [13].

In case of flow condensation inside horizontal tubes vapour enters the tube being in the superheated state and undergoes changes through flow structures. It first forms the annular flow with superheated steam in the core and liquid on the walls. That region is known as the wet wall desuperheating region. In that region a high thermodynamical non-equilibrium is present as superheated vapour and subcooled liquid are present simultaneously. Further heat removal leads to slug and stratified flow (plug and bubbly) structures and finally subcooled liquid, Fig. 1.2.

In theoretical considerations of two-phase flows either homogeneous model or separated flow models are used. In the first model the two-phase flow is treated as a homogenous mixture with averaged parameters (the velocity of gaseous and liquid phases are equal). The most effective description is attained by application of the homogeneous model for bubbly and dispersed flow regimes, which are characterized by a fairly uniform distribution of the dispersed phase in a continuous medium (liquid or vapor flow respectively). This model is also efficient for high pressures when the densities of liquid and vapor phases approach each other. The separated flow model allows for the difference in phase velocities and is efficient in description of stratified (in horizontal channel), annular, wave and other flows.

The main parameters describing vapor-liquid flow are the local two-phase flow multiplier R, void fraction φ (relative volume or cross section of the channel occupied by vapor), quality x ($x = (h - h_L)/h_{LG}$, where h is the flow enthalpy, h_L

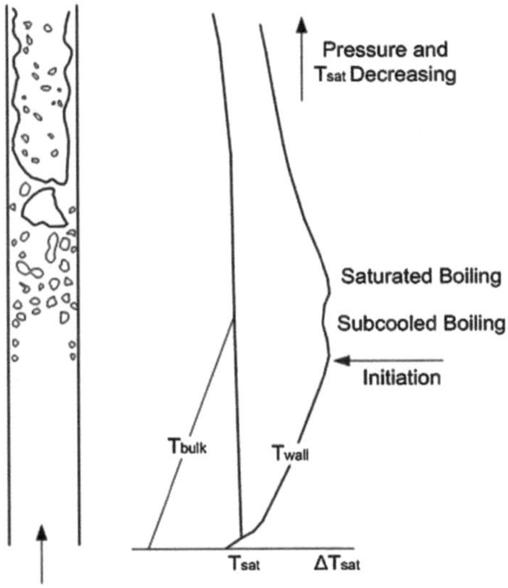

Fig. 1.1 Regions of heat transfer in forced convection

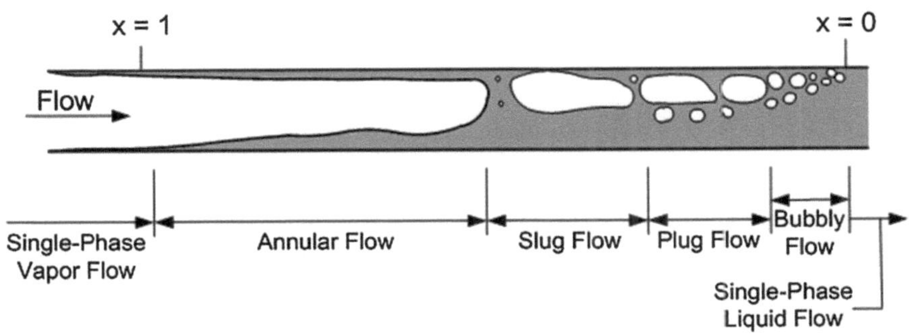

Fig. 1.2 Horizontal co-current flow with condensation

the saturated liquid enthalpy, and h_{LG} the latent heat of vaporization) and the slip ratio s (ratio of mean velocity of vapor to mean velocity of liquid). These three parameters are interconnected through the expression:

$$\frac{x}{1-x} = \frac{\varphi}{1-\varphi} s \frac{\rho_G}{\rho_L} \tag{1.1}$$

where s denotes the slip between the phases: $s = w_G/w_L$.

1.2.1 Pressure drop

The overall pressure drop in the channel Δp_{TP} is the sum of three components, namely the pressure drop due to friction, pressure drop due to acceleration and hydrostatic pressure drop, respectively:

$$\Delta p_{TP} = \Delta p_f + \Delta p_{ac} + \Delta p_h \tag{1.2}$$

In (1.2) $\Delta p_{ac} = \Delta\left[\frac{x^2}{\rho_G \varphi} + \frac{(1-x)^2}{\rho_L(1-\varphi)}\right]$ is the component of pressure drop due to the flow acceleration (of liquid and vapor phases) owing to the change of vapor quality, and $\Delta p_h = \Delta[\rho_L(1-\varphi) + \rho_G \varphi]g$, the pressure drop brought about by overcoming the hydrostatic pressure (in vertical system).

Flow resistance due to friction is greater than in case of single chase flow with the same flow rate. Usual approach is to model the pressure drop due to friction in terms of the two-phase flow multiplier and a single phase flow pressure drop. The two-phase flow multiplier is defined as a ratio of pressure drop in two-phase flow, Δp_f, to the total pressure drop in the flow with either liquid of vapour, Δp_O, present:

$$R = \frac{\Delta p_f}{\Delta p_O} \tag{1.3}$$

Precise determination of void fraction for two-phase flow is crucial for correct evaluation of pressure drop and heat transfer coefficient. Solving (1.1) with respect to void fraction one can obtain:

$$\varphi = \frac{x\frac{\rho_L}{\rho_G}}{x\frac{\rho_L}{\rho_G} + s - sx} \tag{1.4}$$

In literature there exist a number of relations describing void fraction. In the case of homogenous model of two-phase flow the slip is assumed as unity, $s = 1$. In case of the separated model the slip s can be for example determined from the Zivi [14] formula:

$$s = \sqrt[3]{\frac{\rho_L}{\rho_G}} = \sqrt[3]{\eta} \tag{1.5}$$

where $\eta = \rho_L/\rho_G$. Another often used relation for the slip is due to Chisholm:

$$s(x) = [1 - x \cdot (1-\eta)]^{0.5} \tag{1.6}$$

1.2.2 Review of Selected Approaches for Determination of Two-Phase Flow Multipliers

The local two-phase flow multiplier, R, can be evaluated in accordance to the assumed flow model. From amongst existing in literature correlations the models due to Lockhart–Martinelli, Chisholm, Friedel, Muller-Steinhagen and Heck and homogeneous model are advised for use in case of conventional size channels.

Most often used relations for the two-phase flow multiplier R are listed below:

a. Homogeneous model

$$R(x) = \left[\frac{\left(\frac{dp}{dz}\right)_{TP}}{\left(\frac{dp}{dz}\right)_{LO}}\right]^{0.5} = 1 + (\eta - 1)\cdot x \tag{1.7}$$

b. Lockhart–Martinelli model [15]

$$R_{L-M}(x) = \left[\frac{\left(\frac{dp}{dz}\right)_{TP}}{\left(\frac{dp}{dz}\right)_{L}}\right]^{0.5} = \left(1 + \frac{C}{X_{tt}(x)} + \frac{1}{X_{tt}^2(x)}\right)(1-x)^{1.75} \tag{1.8a}$$

where the Martinelli's parameter X_{tt}

$$X_{tt} = \frac{\left(\frac{dp}{dz}\right)_{L}}{\left(\frac{dp}{dz}\right)_{G}} = \left(\frac{1-x}{x}\right)^{0.875}\left(\frac{\rho_g}{\rho_l}\right)^{0.5}\left(\frac{\mu_l}{\mu_g}\right)^{0.125} \tag{1.8b}$$

where $(dp/dz)_L$ and $(dp/dz)_G$ are the frictional pressure gradients for the liquid and gas phase respectively flowing alone in the channel. In (1.8b) the turbulent flow correlations were used. Values of the constant C with respect to the flow regime are shown in Table 1.1.

c. Chisholm method [16]

$$R_{Ch} = \frac{\left(\frac{dp}{dz}\right)_{TP}}{\left(\frac{dp}{dz}\right)_{LO}} = 1 + (Y^2 - 1)\left[Bx^{(2-n)/2}(1-x)^{(2-n)/2} + x^{2-n}\right] \tag{1.9}$$

In relation (1.9) value of exponent $n = 0.25$, whereas B is a function of the mass velocity and Martinelli parameter.

d. Muller-Steinhagen and Heck model [17]

$$R_{MS} = \frac{\left(\frac{dp}{dz}\right)_{TP}}{\left(\frac{dp}{dz}\right)_{LO}} = \left[1 + 2\left(\frac{1}{f_1} - 1\right)\right](1-x)^{1/3} + \frac{1}{f_1}x^3 \tag{1.10}$$

where the function $f_1 = \left(\frac{\mu_L}{\mu_G}\right)^{0.25}\cdot\left(\frac{\rho_L}{\rho_G}\right)^{-1}$.

Table 1.1 Values of constant C from equation (1.8b)

Flow	Liquid	Gas	C
	Turbulent	Turbulent	20
	Laminar	Laminar	5
	Turbulent	Laminar	10
	Laminar	Turbulent	12

e. Friedel [18]

$$R_F = \frac{\left(\frac{dp}{dz}\right)_{TP}}{\left(\frac{dp}{dz}\right)_{LO}} = E + \frac{3.24FH}{Fr^{0.045}We^{0.035}} \tag{1.11a}$$

Expressions for E, F and H present in (1.11a) are defined in the form:

$$\begin{aligned} E &= (1-x)^2 + x^2\left(\frac{\rho_l f_G}{\rho_g f_L}\right) \\ F &= x^{0.78}(1-x)^{0.224} \\ H &= \left(\frac{\rho_L}{\rho_G}\right)^{0.91}\left(\frac{\mu_G}{\mu_L}\right)^{0.19}\left(1-\frac{\mu_G}{\mu_L}\right)^{0.7} \end{aligned} \tag{1.11b}$$

Unfortunately, the correlations developed for conventional size tubes cannot be used in calculations of pressure drop in minichannels. In case of small diameter channels there are other correlations advised for use. Their major modification is the inclusion of the surface tension effect into existing conventional size tube correlations. Amongst the most acknowledged ones are those due to Mishima and Hibiki [19], Tran et al. [20] and Zhang and Webb [21].

Mishima and Hibiki [19] investigated various flow structures, void fraction and rates of bubble increase in slug flow as well as pressure drop of air-water mixtures in vertical tubes of diameter from the range 1.05–3.9 mm and tube length 210–1,000 mm. Their results proved to be consistent with the correlation due to Lockhart–Martinelli (1.8a), where the parameter C was modified to yield:

$$C = 21 \cdot \left(1 - e^{-0.319 d_h}\right) \tag{1.12a}$$

where d_h is the hydraulic diameter and correlation is applicable for circular and square horizontal as well as vertical channels. Recently, the parameter C was further modified for the case where liquid and gas Reynolds numbers are smaller than 2,000 to yield:

$$C = 21 \cdot \left(1 - e^{-0.358/Con}\right) \tag{1.12b}$$

Method is applicable for the range of hydraulic diameters 0.014 mm < d_h < 6.25 mm ($Con = (\sigma/(g(\rho_l - \rho_v)D_h^2)^{0.5}$).

Another well established correlation in literature for small diameter channels is the one due to Tran et al. [20], who established their method based on measurements of diabatic pressure drop of R134a, R12 and R113. Measurements were established based on measurements in two circular channels of internal diameters 2.46 and 2.92 mm and one square channel of cross-section 4.06 × 1.7 mm, for six pressures from the range 138–856 kPa. Authors devised a new correlation, based on the modification of the Chisholm method (1.9), where the surface tension effect was incorporated through the so called constraint number, Con.

$$R = \left[\frac{\left(\frac{dp}{dz}\right)_{TP}}{\left(\frac{dp}{dz}\right)_{LO}}\right]^{0.5} = 1 + \left(CY^2 - 1\right)\left[Con\, x^{0.875}(1-x)^{0.875} + x^{1.75}\right] \tag{1.13}$$

where C = 4.3, $Y^2 = (dp/dz)_{GO}/(dp/dz)_{LO}$. That correlation reveals very well authors own experimental data, as it captures about 93.8 % of measurements within ±20 %.

Zhang and Webb [21] investigated the friction term of pressure drop in horizontal flow of R134a, R22 and R404A in tubes with diameters ranging from 2.16 to 6.20 mm. Authors concluded that the metod of calculation of due to Friedel [18] overestimates their experiments and devised a local two-phase flow multiplier in the form:

$$R_{ZW} = \left[\frac{\left(\frac{dp}{dz}\right)_{TP}}{\left(\frac{dp}{dz}\right)_{LO}}\right]^{0.5} = (1-x)^2 + 2.87 \cdot (x)^2 \cdot \left(\frac{p}{p_{kr}}\right)^{(-1)} + 1.68 \cdot (1-x)^{0.25} \cdot \left(\frac{p}{p_{kr}}\right)^{(-1.64)} \tag{1.14}$$

1.3 Review of Selected Correlations for Flows with Change of Phase

1.3.1 Selected Flow Boiling Correlations

The topic of flow boiling predictions has been scrutinized for over half a century as the interest in that kind of heat transfer started in early sixties. In the present paper it is not the intention of the author to provide a survey of all available methods for that purpose, but only to indicate the major approaches to modeling of flow boiling heat transfer in conventional and small diameter channels. For an extensive literature survey of flow boiling in conventional size channels the reader is referred to Thome [22] or in small diameter channels to Bergles et al. [23], Kandlikar [1] or

Thome [24]. In general, all existing approaches are either the empirical fits to the experimental data, or form an attempt to combine two major influences to heat transfer, namely the convective flow boiling without bubble generation and nucleate boiling. Generally that is done in a linear or non-linear manner. Alternatively, there is a group of modern approaches based on models which start from modeling a specific flow structure and in such way postulate more accurate flow boiling models, usually pertinent to bubbly/slug or annular flows.

The empirical correlations suggested up to date are based on reduction of a restricted number of authors own experimental data or form a generalisation of a greater number of experimental data from various authors. In the latter case correlations are usually of a worsen accuracy in predicting the heat transfer coefficient or the pressure drop, due to the fact that each experiment contributes with its own systematic measurement error. However, such correlations are still of more general character. All such empirical correlations are the fits to experimental data, which restricts their generality. In the case of a lack of generation of bubbles the experimental data in conventional size tubes, i.e. having diameters greater than 3 mm, [1], are usually modeled in function of Martinelli parameter X_{tt}, defined by (1.8b), in the form:

$$\frac{\alpha_{TPB}}{\alpha_{LO}} = a(X_{tt})^b \tag{1.15a}$$

Dengler and Addoms [25] suggested values of parameters a and b to be a = 3.5 and b = −0.5 respectively, whereas Guerrieri and Talty [26] determined them as a = 3.4 and b = −0.45. In the case the bubble generation is present in the flow such approach encounters some limitations and in order to alleviate it an approach based on incorporation of the boiling number, $Bo = q_w/(Gh_{LG})$, into the form of correlation (1.15a) is often used. A general form of heat transfer coefficient with account of bubble generation yields:

$$\frac{\alpha_{TPB}}{\alpha_{LO}} = a'(Bo + mX_{tt})^{b'} \tag{1.15b}$$

Schrock and Grossman [27] recommended the correlation of the type (1.15b) where coefficients a′, b′ and m assume values: a′ = 7400, b′ = 0.66 and m = 0.00015. Collier and Pulling [28] suggested another set of coefficients in (1.15b): a′ = 6700, b′ = 0.66 and m = 0.00035. On the basis of that approach several other correlations for conventional channels have been developed, just to mention those due to Shah [29], Kandlikar [30] and Gungor and Winterton [31].

The most popular approach, however, to model flow boiling is to present the resulting heat transfer coefficient in terms of a combination of nucleate boiling heat transfer coefficient and convective boiling heat transfer coefficient in the following manner:

$$\alpha_{TPB} = \left[(\alpha_{LO}F)^n + (\alpha_{PB}S)^n\right]^{1/n} \tag{1.16}$$

where α_{PB} denotes pool boiling heat transfer coefficient and α_{LO} is the liquid convective heat transfer coefficient, which can be evaluated using for example the Dittus–Boelter type of correlation. Exponent n is an arbitrarily selected natural number without the recourse to any theoretical foundations. Function S is the so called suppression factor which accounts for the fact that together with the increase of vapour flow rate the effect related to forced convection increases, which on the other hand impairs the contribution from nucleate boiling, as the thermal layer is reduced. The parameter F accounts for the increase of convective heat transfer with increase of vapour quality. That parameter always assumes values greater than unity, as flow velocities in two-phase flow are always greater than in the case of single phase flow. The approach represented by Eq. (1.16) is usually dedicated to Rohsenow [32], who suggested a linear superposition with $n = 1$, which has been later modified by Chen [33], who incorporated the suppression and enhancement functions, S and F respectively. The correlation due to Chen is used up to date with a significant appreciation in case of boiling flows in conventional size tubes. Another approach to derive the exponent n in Eq. (1.16) was made by Kutateladze [34], who recommended a superposition approach, but combined in a geometrical rather than linear manner with the value of exponent $n = 2$. A similar summative non-linear approach has been recommended later by Steiner and Taborek [35] with $n = 3$. There is also an issue of the choice of appropriate correlation selection for calculation of pool boiling heat transfer coefficient, as Chen [33] used the model due to Forster and Zuber [36], whereas later studies tend to use rather the more general correlation due to Cooper [37], which enables calculations of pool boiling heat transfer coefficient for different fluids.

Following that brief survey of superposition models the question arises on how to select the appropriate value of exponent n, in equation (1.16). Should there be a value of $n = 1$ or 2, or 3 assumed? Or maybe other value of that exponent ought to be looked for? Mikielewicz [38], provided an answer to that question showing on the basis of consideration of energy dissipation in the two-phase flow that the value of exponent should be $n = 2$. A modified version of that model will be examined in this book.

In case of small diameter channels some empirical correlations found appreciation from users as being superior to the conventional size tubes. Lazarek and Black [39], measured the local and average heat transfer coefficient, pressure drop, and critical heat flux of saturated boiling of R-113 flowing vertically upwards and downwards in 3.17 mm tubes, of the length $L = 123$ and 246 mm, $G = 125$–750 kg/(m^2 s), $p = 1.3$–4.1 bar, $q_w = 14$–380 kW/m^2, and $\Delta T_{sub} = 3$–73 K. The heat transfer coefficient was found to be independent of x for $x \geq 0$. They presented an empirical correlation for local saturated boiling heat transfer coefficients, where the Nusselt number is a function of the liquid Reynolds number and the Boiling number.

$$\alpha_{TPB} = 30(\mathrm{Re}_{LO})^{0.857} Bo^{0.714} \frac{\lambda}{d} \quad (1.17)$$

Steinke and Kandlikar [40] recommend that in the case of Reynolds number smaller than 1,600, i.e. $Re_{LO} < 1,600$, the heat transfer coefficient α_{LO} should be determined from the laminar flow range correlations, such as in case of constant heat flux, $Nu_{LO} = 4.36$. The two-phase flow heat transfer coefficient α_{TPB} is a greater value of the two α_{NBD} or α_{CBD}:

$$\alpha_{NBD} = 0.6683Co^{-0.2}\alpha_{LO} + 1058.0Bo^{0.7}F_{fl}\alpha_{LO} \tag{1.18a}$$

$$\alpha_{CBD} = 1.1360Co^{-0.9}\alpha_{LO} + 667.2Bo^{0.7}F_{fl}\alpha_{LO} \tag{1.18b}$$

The parameter F_{fl} is a fluid dependent factor, describing the pool boiling of particular fluid. Value of F_{fl} can be read from relevant tables for different fluids.

Kattan et al. [41] concluded that only the models based on distinguishing between flow regimes should be genuinely considered for a general use in prediction of heat transfer coefficient in horizontal tubes. A model has been developed by them based on a flow map. Some refinement to that approach was introduced by Wojtan et al. [42]. Such models seem to be promising, however, fall to the class of the regime dependent models. They require prior knowledge of the particular flow regime, as well as the proportion of liquid and gaseous phase in the flow. For that reason such model is difficult to be applied in engineering practice.

A completely different kind of approach to model flow boiling is presented in tackling the problem from the principles of conservation of mass, momentum and energy, and subsequently through a numerical solution to such problem. An example of such approach is a four-field two-fluid model due to Lahey and Drew [43]. In such approach the two phases of fluid can exist in continuous or dispersed form leading to the occurrence of four fields, namely continuous liquid-dispersed gas and continuous gas-dispersed liquid. Some success has been obtained in modeling of bubbly flows and annular flows, however the major challenge is to predict the flow development and transformation through consecutively developing flow structures, Podowski [44]. The closure models and jump conditions form a very difficult task to be implemented into the calculation procedure, similarly as the transient conditions. Apparently the research is still being exercised in that direction but it is still some time to go before useful results are to be obtained of simulation of the entire transformation from subcooled liquid to superheated vapour.

1.3.2 Selected Flow Condensation Correlations

Condensation inside tubes has been a topic of interest of not too many investigations. Mentioned here for example should be studies by Cavallini et al. [45], El Hajal et al. [46], Thome et al. [47] and Garimella [48]. Flow condensation at high heat fluxes enables removal of significant heat fluxes. In case of condensation in small diameter channels the surface phenomena together with the characteristics of the surface itself become more important, as well as interactions between the wall and fluid.

A pioneering work to modeling of flow condensation was presented by Akers et al. [49], valid for the most commonly found flow structure, namely the annular flow. Empirical correlation due to Shah [50] is another of the most widely used methods for calculations of heat transfer coefficients in flow condensation. In the development of that model it was concluded that in the case of lack of nucleate boiling, which is the case for condensation, the heat transfer coefficient should be close to the one for the annular flow structure. Traviss and Rohsenow [51] used the analogy between exchange of heat and momentum using universal velocity distribution to obtain correlation for heat transfer coefficient in the annular flow. On the basis of assumed velocity profile the authors obtained a relation describing the heat transfer coefficient during condensation as a function of turbulent liquid film thickness. Dobson and Chato [52] noticed that the method of analysis of the boundary layer, used by some researchers and in that light by Traviss and Rohsenow [51] in particular, is similar to the approach utilizing the two-phase flow multiplier, used by other authors. They found that at the foundation of thermal resistance in the annular flow are the laminar and buffer sublayers. Hence it is necessary to incorporate the multi-zone model of thermal resistance in liquid film, considering also the presence of waves at the phase interface or variation of liquid film thickness.

In microchannels we observe domination of forces resulting from action of surface tension and viscosity over the gravitational forces. Hence the attempt to extend the range of validity of correlations developed for conventional channels onto the channels with small diameters often leads to errors in pressure drop and heat transfer description, making such approaches useless. Additionally, the heat transfer coefficient and pressure drop in microchannels strongly depend upon the quality, especially in case when annular flow structure is encountered. Hence the detection of flow structures and their influence on pressure drop and heat transfer is indispensable during the condensation of the fluid.

Cavallini et al. [53] proposed separate correlations for the annular, annular-stratified and stratified and slug flow regimes. This method is based upon a large data bank, collected for halogenated refrigerants inside tubes with internal diameter $d > 3$ mm at reduced pressure $p_R < 0.75$ and density ratio (ρ_L/ρ_G) Author use their correlation to predict heat transfer coefficient during condensation inside tubes with internal diameter $d < 3$ mm. The applicable flow regimes were selected based on criteria similar to those proposed by Breber et al. [54], when at the dimensionless vapour velocity $J_G < 2.5$ and $X_{tt} < 1.6$ the flow enters the annular-stratified flow transition and stratified flow region. The heat transfer coefficient $\alpha_{an\text{-}st}$ is calculated from a linear interpolation between heat transfer coefficient at the boundary of the annular flow region $\alpha_{an,JG=2.5}$ and that for fully stratified flow α_{st}. When $X_{tt} > 1.6$ and $J_G < 2.5$, the flow enters the stratified-slug transition and slug flow pattern region, the heat transfer coefficient is calculated as linear interpolation between the coefficient computed at $X_{tt} = 1.6$ and the one for the liquid flowing with the entire flow rate. Model for annular flow is applied when the dimensionless vapour velocity is lower than 2.5. For the annular flow regime,

Cavallini et al. [53] suggested the use of the heat transfer model proposed by Kosky and Staub [55] with the modified Friedel [18] correlation for shear stress:

$$\alpha_{TPK}=\begin{cases}\alpha_{an}=\dfrac{\rho_L c_{pL}\left(\frac{\tau}{\rho_L}\right)^{0.5}}{T^+} & for \quad J_G>2.5\\ \alpha_{an-st}=\left(\alpha_{an,J_G=2.5}-\alpha_{st}\right)\left(\dfrac{J_G}{2.5}\right)+\alpha_{st} & for \quad 1.6<J_G\le 2.5\\ \alpha_{an-lt}=\alpha_{LO}+\dfrac{x(\alpha_{1.6}-\alpha_{LO})}{x_{1.6}} & \end{cases} \tag{1.19}$$

where $J_G=\frac{Gx}{\sqrt{d\,g\,\rho_G(\rho_L-\rho_G)}}$.

The dimensionless film thickness is based on the liquid-phase Reynolds number yields:

$$\delta^+=\begin{cases}\left(\dfrac{\mathrm{Re}_L}{2}\right)^{0.5} & for \quad \mathrm{Re}_L\le 1145\\ 0.0504\mathrm{Re}_L^{7/8} & for \quad \mathrm{Re}_L>1145\end{cases} \tag{1.20}$$

The dimensionless temperature is determined based on the dimensionless film thickness analogously to Traviss et al. [51].

$$T^+=\begin{cases}\delta^+\mathrm{Pr}_L & for \quad \delta^+\le 5\\ 5\left\{\mathrm{Pr}_L+\ln\left[1+\mathrm{Pr}_L\left(\frac{\delta^+}{5}-1\right)\right]\right\} & for \quad 5<\delta^+\le 30\\ 5\left\{\mathrm{Pr}_L+\ln(1+5\mathrm{Pr}_L)+0.495\left(\frac{\delta^+}{30}\right)\right\} & for \quad \delta^+>30\end{cases} \tag{1.21}$$

Thome et al. [47] developed a multi-regime heat transfer correlation, in which the regimes identified are either as (a) fully annular forced convective, or as (b) consisting of varying combinations of upper gravity driven, and lower forced convective terms, in case of horizontal flows. Thome et al. [47] proposed heat transfer model for evaporation inside horizontal tubes. They founded that there is a close similarity between the convection mechanisms in annular film condensation and annular film evaporation inside tubes. Finally they proposed the new following model for the annular flow:

$$\alpha_{TPK}=C\mathrm{Re}_L^n\,\mathrm{Pr}_L^m\frac{\lambda_l}{\delta}f_i \tag{1.22}$$

where δ is the film thickness determined using the authors own void fraction model, where f_i reads:

$$f_i=1+\left(\frac{u_G}{u_L}\right)^{0.5}\left[\frac{(\rho_L\rho_G)g\delta^2}{\sigma}\right]^{0.25} \tag{1.23}$$

The constants c, n and m were specified to be 0.003, 0.74 and 0.5, respectively, based on the best fit to experimental data for tubes with $d_h > 3$ mm. This correlation predicts heat transfer coefficients with the average deviation of −74 %. Thome et al. [47] proposed also the general expression for the local condensation heat transfer coefficient, but this method is determined with the aid of a complicated procedure which requires identification of the flow pattern.

The accuracy of predictions using the methods presented above is not fully satisfactory. One of the possible reasons for underestimation of data is that the models based on the annular flow structure are derived using the stresses determined cases for the conventional size channels. There are no counterpart methods for calculation of heat transfer coefficient in small diameter tubes. For that reason most of the approaches to date use empirical correlations for calculations.

1.4 Dissipation Based Model for Pressure Drop and Heat Transfer in Flow Boiling and Flow Condensation

Presented below, in brief, is the model due to Mikielewicz [38], which, as mentioned earlier, has been developed for the case of flow boiling on the basis of consideration of dissipation energy in the flow and recently modified to its final form by Mikielewicz et al. [6, 56]. Modifications incorporated into the model are envisaged to suit predictions both in flow condensation and flow boiling as well as minichannels and conventional channels.

The fundamental hypothesis in the model under scrutiny here is the fact that heat transfer in flow boiling with bubble generation, regarded here as an equivalent flow of liquid with the properties of a two-phase flow, can be modeled as a sum of two contributions leading to the total energy dissipation in the flow, namely the energy dissipation due to shearing flow without the bubbles, E_{TP}, and dissipation resulting from the bubble generation, E_{PB}, Mikielewicz [38]:

$$E_{TPB} = E_{TP} + E_{PB} \tag{1.24}$$

Energy dissipation under steady state conditions in the two-phase flow can be approximated as energy dissipation in the laminar boundary sub-layer, which dominates in heat and momentum transfer in the considered process. Dissipation energy is expressed as power lost in the control volume. The term power refers to compensation of two-phase flow friction losses and is expressed through a product of shear stress and flow velocity. More on that issue can be found in [38]. Analogically can be expressed the energy dissipation due to bubble generation in the two-phase flow. Substituting the definition of respective energies into Eq. (1.24) a geometrical relation between the friction factor in two-phase flow is obtained which forms a geometrical sum of two contributions, namely the friction factor due to the shearing flow without bubbles and the friction factor due to generation of bubbles, in the form:

$$\xi_{TPB}^2 = \xi_{TP}^2 + \xi_{PB}^2 \tag{1.25}$$

It may be difficult to imagine the flow resistance during the generation of bubbles, however in Russian literature such issue has already been scrutinized and there is a number of contributions, where the flow resistance caused merely by generation of bubbles on the wall were reported, see for example Ananiev [57]. In the next section that term will be further scrutinized for the inclusion of non-adiabatic effects. Expression (1.25) forms an underlying hypothesis for determination of heat transfer coefficient in flow boiling and later flow condensation. It enables subsequent derivation of the heat transfer model for a two-phase flow with phase change such as flow boiling and flow condensation.

1.4.1 Heat Transfer Coefficient in Flow Boiling and Flow Condensation

Making use of the analogy between exchange of momentum and heat we can generalize the expression (1.25) to extend the postulated hypothesis over to heat transfer coefficients to yield the appropriate heat transfer coefficient in flow boiling with bubble generation in terms of simpler modes of heat transfer, namely heat transfer coefficient in shearing flow without bubble generation and heat transfer coefficient in nucleate boiling:

$$\alpha_{TPB}^2 = \alpha_{TP}^2 + \alpha_{PB}^2 \tag{1.26}$$

Comparing Eq. (1.26) with Eq. (1.16) we can notice that indeed the value of exponent n = 2. The model postulated here explains therefore on physical grounds, why in Eq. (1.16) there should be used that specific value of exponent n and not any other. Having obtained the geometric summation relation between the heat transfer coefficients we can now evaluate constituent terms in (1.26).

Heat transfer without bubble generation, α_{TP}, which is applicable both to flow boiling and flow condensation, can be modeled in terms of the two-phase flow multiplier. From the definition of the two-phase flow multiplier the pressure drop in adiabatic two-phase flow can be related to the pressure drop of the flow where only liquid at a flow rate G is present:

$$\Delta p_{TP} = R\,\Delta p_{LO} \tag{1.27}$$

The pressure drop in the two-phase flow without bubble generation can also be considered as a pressure drop in the equivalent flow of a fluid flowing with velocity w_{TP}:

$$\Delta p_{TP} = \frac{L}{d}\xi_{TP}\rho_L\frac{w_{TP}^2}{2} \tag{1.28}$$

The pressure drop of the liquid flowing alone can be determined from a corresponding single phase flow relation:

$$\Delta p_{LO} = \frac{L}{d} \xi_{LO} \rho_L \frac{w_{LO}^2}{2} \tag{1.29}$$

In case of turbulent flow we will use the Blasius equation for determination of the friction factor, whereas in case of laminar flow the friction factor can be evaluated from the corresponding expression valid in the laminar regime. In effect obtained is a relation enabling calculation of heat transfer coefficient in flow boiling without bubble generation in the form:

$$\frac{\alpha_{TPB}}{\alpha_{LO}} = \sqrt{R^n} \tag{1.30}$$

That correlation will also be used later in calculations of condensation inside tubes. In (1.30) n = 2 for laminar flows, whereas for turbulent flows the value of exponent n is taking up a value of n = 0.76. A critical difference of that method in comparison to other authors models is incorporation of the two-phase flow multiplier into modelling of convective boiling. In case of refrigerants the two-phase flow multiplier R_{MS} due to Müller-Steinhagen and Heck (1.10) is recommended for use in case of refrigerants, Ould Didi et al. [58] and Sun and Mishima [59]. A different model however should be selected in case of dealing with other fluids.

1.4.2 General Method for Calculation of Heat Transfer in Two-Phase Flows with Change of Phase

Following the derivation presented above the general expression for calculation of heat transfer coefficient in flow boiling and flow condensation can be devised in the form:

$$\frac{\alpha_{TPB}}{\alpha_{LO}} = \sqrt{(R)^n + \frac{C}{1+P}\left(\frac{\alpha_{PB}}{\alpha_{LO}}\right)^2} \tag{1.31}$$

In Eq. (1.31) C = 1 in case of flow boiling modeling and C = 0 for flow condensation.

In (1.31) the term P $= 2.53 \times 10^{-3} \mathrm{Re}_{LO}^{1.17} Bo^{0.6} (R_{MS} - 1)^{-0.65}$ has been developed by a method of multiple regression fitting. The pool boiling heat transfer coefficient α_{PB}, is recommended to be calculated from the Cooper relation [37]. The applied heat flux is incorporated through the boiling number Bo, defined as, Bo = $q/(Gh_{LG})$. For the same difference between the wall and saturation temperature there is a different temperature gradient in the fluid in case of pool boiling and flow boiling. In the case of flow boiling the boundary layer in such case is thinner and hence the gradient of temperature is more pronounced, which suppresses the generation of bubbles. That is the reason why heat flux influence is included in modeling. That term is more important for conventional size tubes, but cannot be totally neglected in case of small diameter tubes in the bubbly flow regime, where it is highly important. In correction P, present in (1.31), the two-phase flow multiplier due to Muller-Steinhagen and Heck (1.10) [17] R_{MS} must be used, as that model was utilised in the procedure of data reduction and determination of the regression coefficients. Here it acts in the correction P as a sort of the convective number.

Care was taken that the expression (1.31) reduces to liquid only heat transfer coefficient in case when quality $x = 0$ and vapour only heat transfer coefficient if $x = 1$. In order to attain that the two-phase flow multiplier model due to Muller-Steinhagen and Heck [17] was modified to incorporate the function f_{1z}, which secures such limiting behavior. The resulting expression now reads:

$$R_{MS} = \left[1 + 2\left(\frac{1}{f_1} - 1\right)x\right] \cdot (1 - x)^{1/3} + x^3 \frac{1}{f_{1z}} \tag{1.32}$$

In Eq. (1.32) $f_1 = (\rho_L/\rho_G)(\mu_L/\mu_G)^{0.25}$ for turbulent flow and $f_1 = (\rho_L/\rho_G)(\mu_L/\mu_G)$ for laminar flows. Introduction of the function f_{1z}, expressing the ratio of heat transfer coefficient for liquid only flow to the heat transfer coefficient for gas only flow, is to meet the limiting conditions, i.e. for $x = 0$ the correlation should reduce to a value of heat transfer coefficient for liquid, $\alpha_{TPB} = \alpha_L$ whereas for $x = 1$, approximately that for vapour, i.e. $\alpha_{TPB} \cong \alpha_G$. Hence:

$$f_{1z} = \frac{\alpha_G}{\alpha_L} \tag{1.33}$$

where $f_{1z} = (\lambda_G/\lambda_L)$ for laminar flows and for turbulent flows $f_{1z} = (\mu_G/\mu_L)(\lambda_L/\lambda_G)^{1.5}(c_{pL}/c_{pG})$. The applicability of correlation (1.31) seems to be quite general, as confirmed for example by the study by Chiou et al. [60] with the fluids not considered in derivation of the model constants.

It is a major drawback of most correlations developed for calculation of heat transfer in conventional size tubes that their accuracy significantly drops when applied to small size tubes, regarded here as smaller than 3 mm, [1]. In such situation the surface tension effects become to be more dominant and need to be reflected in the model. Most of experimental data indicate that most important for small channels is the convective flow boiling mode which, as stems from (1.31), is dependent on the flow resistance. In such case a great deal of care must be exercised to use the appropriate definition of the two-phase flow multiplier model. Again, available empirical correlations of two-phase flow friction fail to be accurate for small diameter channels. For that reason the Muller-Steinhagen and Heck two-phase multiplier correlation was modified to incorporate the additional term responsible for surface tension effects, namely the constraint number Con, [2].

$$Con = \left\{ \frac{\sigma}{[g(\rho_l - \rho_v)]D_h^2} \right\}^{0.5} \tag{1.34}$$

In the form applicable to conventional and small diameter channels the modified Muller-Steinhagen and Heck model described by Eq. (1.10) yields:

$$R_{MS} = \left[1 + 2\left(\frac{1}{f_1} - 1\right) x\, Con^m \right] \cdot (1 - x)^{1/3} + x^3 \frac{1}{f_1} \tag{1.35}$$

where m = 0 for conventional channels. Best consistency with experimental data, in case of small diameter and minichannels, is obtained for m = −1.

1.4.3 Non-Adiabatic Effects in Flow Boiling and Flow Condensation

Analogy between the momentum exchange and heat transfer is by no means a reversible one, especially in case of flows with phase change There are specific effects related to the shear stress modifications, named here the non-adiabatic effects, which will be described below.

1.4.3.1 Non-Adiabatic Effects in Annular Flow

The shear stress between vapour phase and liquid phase is generally a function of non-adiabatic effects. That is a major reason why up to date approaches, considering the issue of flow boiling and flow condensation as symmetric phenomena, are failing in that respect. The way forward is to incorporate a mechanism into the convective boiling term responsible for modification of shear stresses at the vapour-liquid interface. As our final objective is to devise a heat transfer model applicable both to flow boiling and flow condensation we will attempt to improve the shear stress between liquid and vapour phase in annular flow by incorporation of the so called "blowing parameter", B, which contributes to the liquid film thickening in case of flow condensation and thinning in case of flow boiling, Mikielewicz [61]. The model, however, was developed on the basis of injection of gas bubbles into the boundary layer. The migration of bubbles, however, can also be modelled in the similar manner. The devised in [61] formula for modification of shear stresses in the boundary layer reads:

$$\tau^+ = 1 + \frac{B}{\tau_0^+} u^+ \tag{1.36}$$

In Eq. (1.36) $\tau^+ = \tau/\tau_w$, $\tau_0^+ = \tau_w/\tau_{w0}$, where τ_{w0} is the wall shear stress in case where the non-adiabatic effects are not considered, and $B = 2\vartheta_0/(c_f\, u_\infty)$ is the so called “blowing parameter”. Additionally, ϑ_0 denotes the transverse velocity, which in case of condensation or boiling is equal to $q_w/(h_{LG}\,\rho_L)$. In case when Re → ∞ the solution of Eq. (1.36) tends to that suggested by Kutateladze and Leontiev:

$$\tau_0^+ = \left(1 - \frac{B}{4}\right)^2 \tag{1.37}$$

On the other hand, in case of small values of B the solution given by Eq. (1.37) reduces to that recommended by Wallis [63]:

$$\tau_0^+ = \left(1 - \frac{B}{2}\right) \tag{1.38}$$

The analyses due to Kutateladze and Leontiev [62] and Wallis [63] were carried out for the case of flow boiling.

There are also studies in literature, where the shear stress modification in flow condensation is considered, Bai et al. [64]. These authors recommended to consider stresses due to interaction between liquid and vapour phases which are algebraically summed up. The first term is the frictional shear stress due to different velocities of the liquid and vapour phases, whereas the second one is the momentum-transfer shear stress due to vapour condensation:

$$\tau_0 = \tau_f + \tau_{mt} \tag{1.39}$$

Shear stress due to different velocities of vapor and liquid phases is:

$$\tau_f = \pm\frac{1}{2} f \rho_G (u_G - u_L)^2 \tag{1.40}$$

The “+” sign is used when the average velocity of vapour phase is greater than that of liquid at the liquid/vapour interface, otherwise “−” is applied. The momentum-transfer term to vapour condensation:

$$\tau_{mt} = \frac{q_w}{h_{LG}}(u_G - u_L) \tag{1.41}$$

Following incorporation of the above contributions the modified shear stress at the vapour-liquid interface yields:

$$\tau_0^+ = \frac{\tau_0}{\tau_f} = \left(1 + \frac{\frac{q_w}{h_{LG}}(u_G - u_L)}{\pm\frac{1}{2} f \rho_G (u_G - u_L)^2}\right) = \left(1 \pm \frac{B}{2}\right) \tag{1.42}$$

The general form of Eq. (1.42) reminds closely that of Eq. (1.38), confirming that the shear stresses at the liquid/vapour interface can be modified also in case of flow condensation (in such case the “+” sign is applicable). In the present work blowing parameter is:

$$B = \frac{2\vartheta_0}{c_f u_\infty} = \frac{2q}{c_{f0}(u_G - u_L)h_{LG}\rho_G} = \frac{2q\frac{\rho_L}{\rho_G}}{c_{f0}G(s-1)h_{LG}} \tag{1.43}$$

The considerations by Mikielewicz [61] were pertaining to both cases, i.e. flow boiling and flow condensation. Having acquired the way to modify the stresses in flow boiling and flow condensation it is relatively straightforward to implement Eq. (1.36) in the model of pressure drop, which will be presented in next subsection, [56].

1.4.3.2 Non-Adiabatic Effects in Other than Annular Flows

Flow resistance under non-adiabatic conditions differs from the adiabatic case. Majority of research into flow resistance, however, was related to adiabatic flow conditions as investigations were accomplished for air-water mixtures. Only a few studies have been devoted to vapour-liquid mixtures where the value of applied heat flux is important on the extent of flow resistance. In case of annular flow structure the influence of phase change was modeled using the so called "blowing parameter", as presented in previous section. Such approach is, however, not possible in case of dealing with the bubbly flow structure or droplets flow. Presented below is author's approach to incorporate the fact of non-adiabatic effects into other than annular structures. Expressing the flow resistance of a two-phase adiabatic flow through the two-phase flow multiplier in the form:

$$\xi_{TP} = R\xi_{TP\,ad} \tag{1.44}$$

On the basis of the thermal-hydraulic analogy we can assume that the friction factor for pool boiling is in proportion to the pool-boiling heat transfer coefficient, which therefore yields:

$$\xi_{PB} \sim \alpha_{PB} \tag{1.45}$$

That approach is justified through the Stanton analogy originally postulated in the form:

$$Nu = \frac{\alpha d}{\lambda} = \frac{c_f}{2}\mathrm{Re}\,\mathrm{Pr} = \frac{\xi}{8}\mathrm{Re}\,\mathrm{Pr} \tag{1.46}$$

The pool-boiling heat transfer coefficient, which is expressed as a function of wall heat flux, has the form, Cooper [37]:

$$\alpha_{PB} = 55p_r^{0.12}(-\log p_r)^{-0.55}M^{-0.5}(q_w)^{2/3} = C\,q_w^{2/3} \tag{1.47}$$

Substituting the result (1.47) into (1.25)

$$\xi_{TPB}^2 = R^2 \xi_0^2 + \left(\frac{8\alpha_{PB} d}{\lambda \mathrm{Re}\,\mathrm{Pr}} \right)^2 \tag{1.48}$$

After re-arranging of (1.48) we can write the two-phase flow multiplier, which incorporates the non-adiabatic effect in the form:

$$R_{TPB} = \frac{\xi_{TPB}}{\xi_0} = \sqrt{R^2 + \frac{\xi_{PB}^2}{\xi_0^2}} = R\sqrt{1 + \frac{\left(\frac{8\alpha_{PB} d}{\lambda \mathrm{Re}\,\mathrm{Pr}} \right)^2}{\xi_0^2 R^2}} \tag{1.49}$$

The two-phase flow multiplier presented by the above equation reduces to adiabatic formulation in case when the applied wall heat flux is tending to zero. Sample predictions showing the performance of Eq. (1.49) are presented in Fig. 1.3.

As can be seen the modification is only pronouncing itself for small qualities. At quality x = 0 the modification renders more than double increase of two-phase flow multiplier and it almost disappears at quality greater than x = 0.1.

1.4.4 General Method for Calculation of Friction Pressure Drop in Two-Phase Flows with Phase Change

Generalising the obtained above results it can be said that the two-phase flow multiplier inclusive of non-adiabatic effects can be calculated, depending upon the particular flow case and the flow structure in the following way:

$$R_{TPB} = \frac{\xi_{TPB}}{\xi_0} = \begin{cases} R\left(1 \pm \frac{B}{2}\right) & \textit{for annular flow boiling and condensation} \\ R\sqrt{1 + \left(\frac{8\alpha_{PB} d}{\lambda \mathrm{Re}\,\mathrm{Pr}\,\xi_0 R}\right)^2} & \textit{for other flow structures} \end{cases} \tag{1.50}$$

In Eq. (1.50) there is no specification of which two-phase flow multiplier model should be applied. That issue is dependent upon the type of considered fluid. The definition of the two-phase flow multiplier with correction (1.50) should be used in Eq. (1.31).

Effect of incorporation of the blowing parameter on pressure drop predictions is shown in Figs. 1.4, 1.5, 1.6 and 1.7. In the presented case the effect of considering the blowing parameter may reach even 20 % effect. In the present calculations a constant value of the blowing parameter is used, however it is envisaged that the distribution of it should result in even greater accuracy of calculations. Such studies are at the moment underway. The new concept is based on the mathematical model of annular flow.

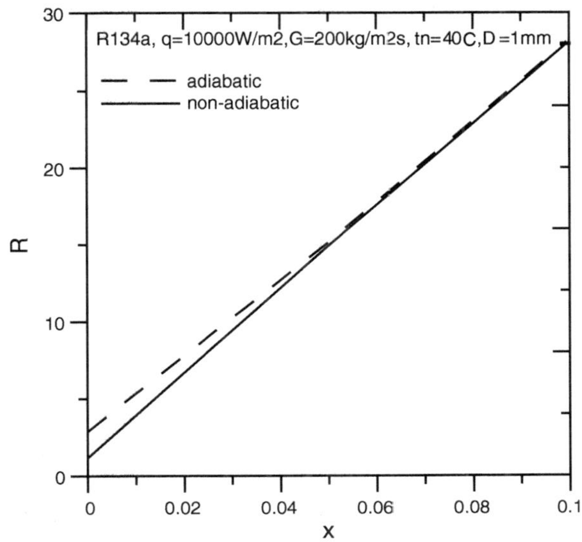

Fig. 1.3 Sample prediction of non-adiabatic effect in bubbly flow

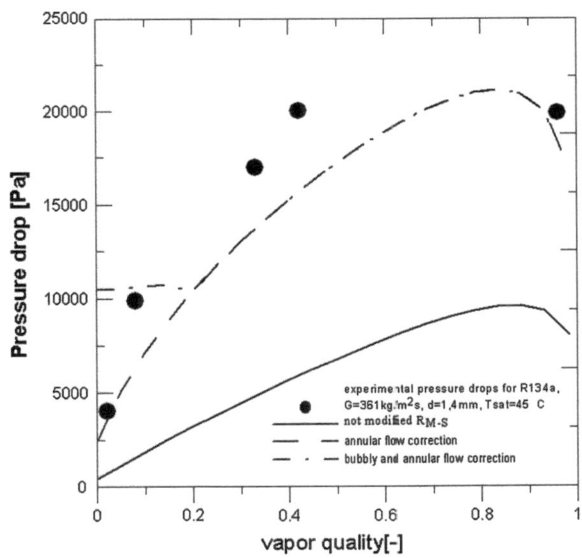

Fig. 1.4 Flow condensation pressure drop distribution in function of quality, R134a, G = 361 kg/m^2 s, d = 1.4 mm, Tsat = 45 °C, [65]

1.4.5 Application of the Model to Subcooled Flow Boiling

Some effort has also been exercised to extend the correlation (1.31) into the subcooled flow regime. In such case we assume that the two-phase flow multiplier tends to some specified value which is close to unity, $R \rightarrow R_{sub}$, as well as the

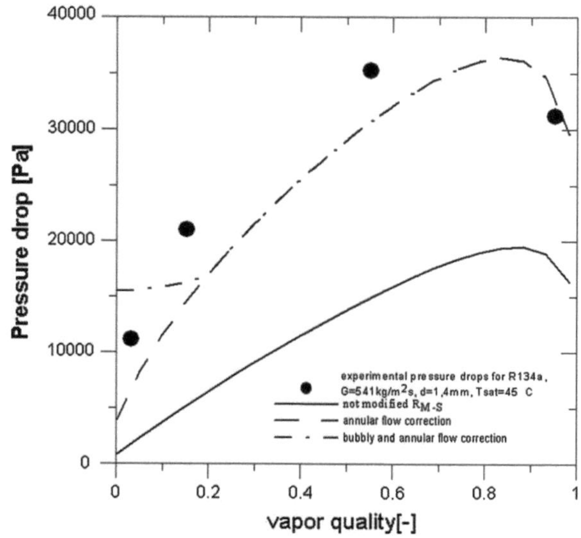

Fig. 1.5 Flow condensation pressure drop distribution in function of quality, R134a, G = 541 kg/m^2 s, d = 1.4 mm, Tsat = 45 °C, [65]

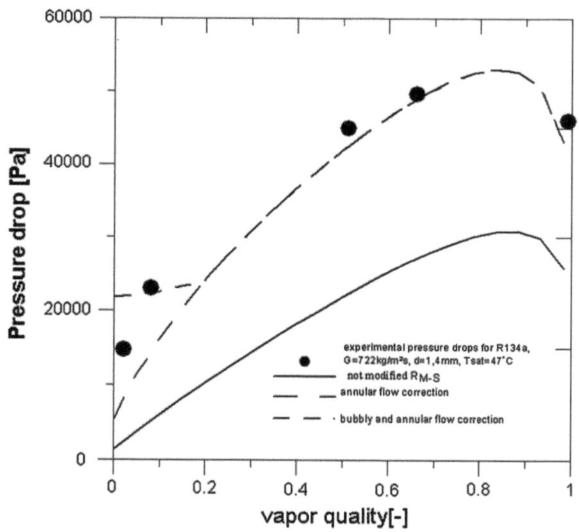

Fig. 1.6 Flow condensation pressure drop distribution in function of quality, R134a, G = 722 kg/m^2 s, d = 1.4 mm, Tsat = 47 °C, [65]

correction function P tends to zero, P → 0. The nucleate boiling heat transfer coefficient in subcooled boiling is related to the Δt_{sat} temperature drop in the channel, instead of total $\Delta t_{sat} + \Delta t_{sub}$, where Δt_{sat} is the difference between the wall temperature and the saturation temperature (superheating), and Δt_{sub} is the difference between the saturation temperature and the mean fluid temperature along the channel (subcooling). In such case the heat transfer coefficient in subcooled boiling is reduced to:

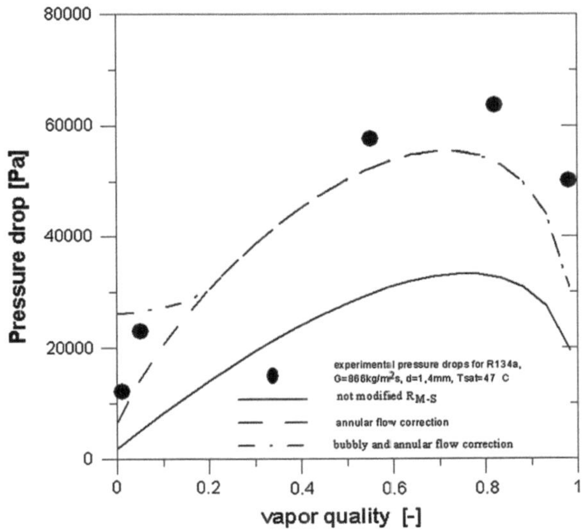

Fig. 1.7 Flow condensation pressure drop distribution in function of quality, R134a, G = 866 kg/m^2 s, d = 1.4 mm, Tsat = 47 °C, [65]

$$\left(\frac{\alpha_{PB_{sub}}}{\alpha_{PB}}\right) = \frac{\Delta t_{sat}}{(\Delta t_{sat} + \Delta t_{sub})} \tag{1.51}$$

If all the above changes are introduced into (1.46) we obtain a new form of correlation for subcooled boiling in the form:

$$\frac{\alpha_{TPB}}{\alpha_{LO}} = \sqrt{R_{sub}^{n} + \left(\left(\frac{\alpha_{PB}}{\alpha_{LO}}\right)\frac{\Delta t_{sat}}{(\Delta t_{sat} + \Delta t_{sub})}\right)^{2}} \tag{1.52}$$

The form of correlation (1.52) can be used for comparisons with some experimental data. Some care is required in specification of the two-phase flow multiplier in that region. To author's knowledge there are no such models in literature, which additionally points at the necessity of further research in that area. Author accomplished some modeling in that area with the experimental data [6]. In that case the best results were obtained for a constant value of $R_{sub} \rightarrow 3$.

1.5 Modelling of Dryout in Minichannels

Dryout condition represents breaking of the continuous liquid contact with the heated surface in a flow channel at moderate to high qualities. It follows the gradual decrease of liquid fraction due to evaporation or boiling of the liquid film. At this point, the liquid film at the heated surface dries up, while entrained droplets may still flow in the vapor core. These droplets occasionally hit the heated surface and evaporate. Dryout occurs, independently of the heat flux, at high vapor qualities. The flow pattern before the dryout condition is usually annular with the vapor phase occupying most of the channel core while the liquid flows as a thin film along the channel wall, Collier and Thome [67].

To simulate the dryout location the original five-equation model was developed. It is based on the solution to the mass conservation balance of liquid in the film and in the core and also the vapor balance in the core of the flow. These equations are supplemented by equations of momentum balance in liquid film and vapour core. The model is dedicated to the dryout phenomena modeling applicable in the conventional and mini-channels. It is an alternative approach to other models proposed so far, see for example Qu and Mudawar [68] and Revellin and Thome [69]. In the case of flow in the small diameter channels it was assumed that:

- conditions of the onset of annular flow are known,
- total balance of mass flux density in the flow is equal to $G = G_f + G_c + G_G$,
- liquid film flow is laminar,
- liquid-vapor interface is stable, therefore the influence of surface tension is large,
- in case of minichannels there is no entrainment of droplets G_E, what was also suggested by Qu and Mudawar [68].

Figure 1.8 illustrates the general hydrodynamic picture of annular flow in two phase microchannel heat tube.

The simplified geometrical representation of annular flow pattern illustrated in Fig. 1.8 denotes the continuous liquid film flowing on the wall, having a laminar character. The liquid film exchanges mass with the two-phase core by means of surface evaporation of the film and deposition of droplets from the core. The liquid film mass velocity is G_f, the core mass velocity is G_c, the gas mass velocity G_G in the core, film thickness, δ, interfacial shear stress, τ_i, and pressure gradient $-(dp/dz)$, constitute the five primary parameters for the annular two-phase flow model. Equations relating the primary parameters are established by application of fundamental conservation equations to both the liquid film and vapor core.

1.5.1 Balance Equations of Mass and Momentum

The model consists of three mass balance equations, namely for liquid in the film as well as two-phase core and the gas phase itself, respectively. These equations are supplemented by the corresponding momentum equations.

1.5.1.1 Mass Balance Equation of Liquid in the Core

$$dG_c \frac{\pi(d-2\delta)^2}{4} = \left(\frac{q_i}{h_{LG}} - G_D\right) \cdot \pi D\, dz \tag{1.53a}$$

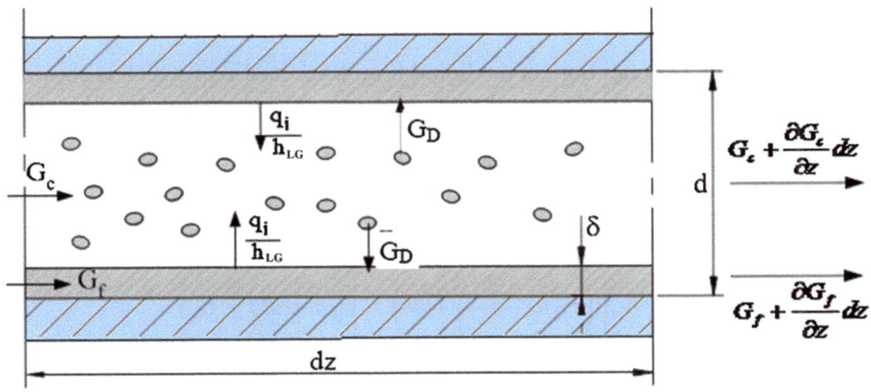

Fig. 1.8 Model of annular flow with boiling

which reads after re-arrangements:

$$\frac{dG_c}{dz} = \frac{4}{d}\left(\frac{q_i}{h_{LG}} - G_D\right) \cdot \frac{1}{\left(1 - \frac{2 \cdot \delta}{d}\right)^2} \tag{1.53b}$$

Mass balance equation of liquid in the film

$$dG_f \frac{\pi\left[d^2 - (d - 2\delta)^2\right]}{4} = \left(G_D - \frac{q_i}{h_{LG}}\right) \cdot \pi\, d\, dz \tag{1.54a}$$

which after re-arrangements yields:

$$\frac{dG_f}{dz} = \frac{4}{d}\left(G_D - \frac{q_i}{h_{LG}}\right) \cdot \frac{1}{1 - \left(1 - \frac{2 \cdot \delta}{d}\right)^2} \tag{1.54b}$$

1.5.1.2 Mass Balance Equation of Gas in the Core

$$dG_G \frac{\pi(d - 2\delta)^2}{4} = \frac{q_i}{h_{LG}} \cdot \pi\, d\, dz \tag{1.55a}$$

which reads after re-arrangements:

$$\frac{dG_G}{dz} = \frac{4}{d}\frac{q_i}{h_{LG}} \cdot \frac{1}{\left(1 - \frac{2\delta}{d}\right)^2} \tag{1.55b}$$

1.5.1.3 Momentum Balance Equation of Liquid in the Film

Let us first focus our attention on the liquid film, a schematic of which is shown in Fig. 1.9. Subscript f present in below equation is equivalent to "L" from nomenclature.

The force balance on the element of film yields:

$$\begin{aligned}&\left[p-\left(p+\frac{\partial p}{\partial z}dz\right)\right]\pi d\delta+(\tau_i-\tau_w)\pi dz\\&+\left[\left(G_f u_f+\frac{\partial(G_f u_f)}{\partial z}\right)-G_f u_f\right]\pi d\delta=0\end{aligned} \tag{1.56a}$$

which can be reduced to the form:

$$-\frac{\partial p}{\partial z}+\frac{\tau_i-\tau_w}{\delta}+\frac{d}{dz}\left(\frac{G_f^2}{\rho_f}\right)=0 \tag{1.56b}$$

1.5.1.4 Momentum Balance Equation of Gas and Liquid in the Core

Next, we consider the core of the flow, Fig. 1.10.

The force balance on the element of core yields:

$$\begin{aligned}&\left[p-\left(p+\frac{\partial p}{\partial z}dz\right)\right]\frac{\pi(d-2\delta)^2}{4}-\tau_i\pi(d-2\delta)dz\\&+\left[\left(G_c u_c+\frac{\partial(G_c u_c)}{\partial z}dz\right)-G_c u_c\right]\frac{\pi(d-2\delta)^2}{4}=0\end{aligned} \tag{1.57}$$

which can be reduced to the form:

$$-\frac{\partial p}{\partial z}-\frac{4\tau_i}{(d-2\delta)}+\frac{\partial}{\partial z}\left(\frac{G_c^2}{\rho_c}\right)=0 \tag{1.58}$$

1.5.2 Determination of Film Thickness

Determination of film thickness from Eqs. (1.53a) ÷ (1.58) enables to calculate the dryout location. As we are primarily interested in that phenomenon, we focus now our attention on determination of the liquid film thickness, which is decreasing in that process. In order to do that we need to scrutinize the momentum equations, as we subtract the equation for the film from the one for the core. That eliminates for us the pressure gradient present in both equations. After such re-arrangements we obtain:

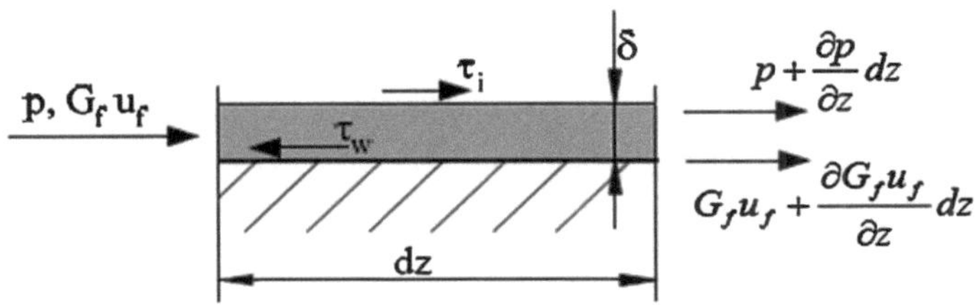

Fig. 1.9 Schematic of forces acting on the film

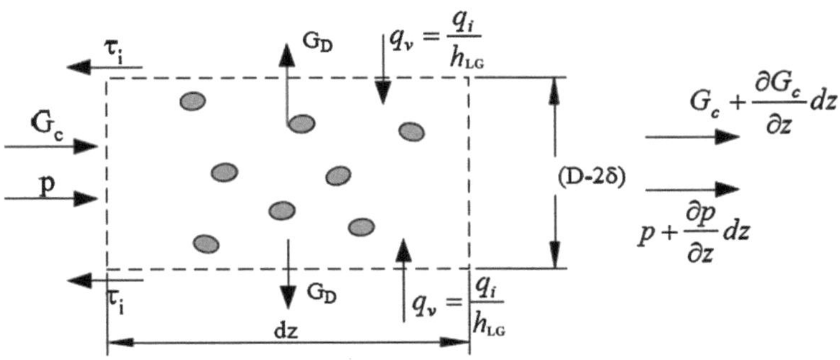

Fig. 1.10 Schematic of the forces acting on the core of the flow

$$\frac{d}{dz}\left(\frac{G_c^2}{\rho_c} - \frac{G_f^2}{\rho_f}\right) = \frac{\tau_i - \tau_w}{\delta} + \frac{4 \cdot \tau_i}{(d - 2\delta)} \tag{1.59}$$

Equation (1.59) solved for the unknown film thickness distribution is:

$$1 + \frac{4\delta}{d - 2\delta} = \frac{\tau_w}{\tau_i} - \frac{\delta}{\tau_i}\frac{d}{dz}\left(\frac{G_c^2}{\rho_c} - \frac{G_f^2}{\rho_f}\right) \tag{1.60}$$

In the next section a brief survey of closure equations, required to be known in calculations is presented.

1.5.3 Closure Equations and Working Equations

The total mass velocity G in the micro-channel is the sum of mass velocities of the liquid film G_f, liquid droplet G_k, and vapor core G_G.

$$G = G_f + G_k + G_G \tag{1.61}$$

1.5.3.1 Heat Flux Distribution

The term related to evaporation of liquid film, which contains the heat flux has been assumed as recommended by Glinski [70] that only a fraction of that applied to the wall takes part in the evaporation process. Till now the wall heat flux was considered as total input present in evaporation process:

$$q_i = q_w \tag{1.62a}$$

Corrected equation for heat flux distribution across the film yields:

$$q_i = q_w \left[\frac{2\delta}{D}\right] \tag{1.62b}$$

1.5.3.2 Location of the Beginning of Annular Flow

The criterion for the transition to annular flow in tubes as proposed by Taitel and Dukler was used to determine the location of the onset of annular flow. This criterion uses a constant value for the Martinelli parameter at the transition point. $X_{ll} = 1.6$, similarly as in Qu and Mudawar [68] where the Martinelli parameter for the combination of laminar liquid and laminar vapor flow rate is given by:

$$X_{ll} = \left(\frac{\mu_L}{\mu_G} \cdot \left(\frac{1 - x_0}{x_0}\right) \cdot \frac{v_L}{v_G}\right)^{0.5} \tag{1.63}$$

Obtained from (1.63) value of vapour quality is used in determination of distribution of liquid between the liquid film and vapor core. The total mass velocity is used for that purpose with following relations:

Amount of droplets at the beginning of annular flow

$$G_k = xG + (0.951 - 0.15 \cdot \sqrt{We}) \tag{1.64}$$

Amount of gas at the beginning of annular flow

$$G_G = G \cdot x \tag{1.65}$$

Mass balance in the core

$$G_c = G_k + G_G \tag{1.66}$$

Amount of liquid in the liquid film

$$G_f = G - G_c \tag{1.67}$$

The Weber number We is defined as $We = G^2 d/(\rho_L \sigma)$.

Deposition of droplets

Denoting the concentration of droplets in the gas core by C and the deposition mass transfer coefficient by k_D, the rate of droplet deposition, G_D is defined by the relation:

$$G_D = k_D \cdot C \tag{1.68}$$

In the present research author relied on empirical methods to determine the deposition mass transfer coefficient, k_D, due to the complexity of this phenomenon, using the correlation:

$$\frac{k_D}{J_G} = 0.009 \cdot \left(\frac{C}{\rho_G}\right)^{-0.5} \cdot \mathrm{Re}_G^{-0.2} \cdot Sc^{-\frac{2}{3}} \tag{1.69}$$

In (1.69) $J_G = G_G/\rho_G$ is vapor superficial velocity, C is the liquid droplet concentration in the vapor core and Bo—the Boiling number $Bo = \frac{q_i}{Gh_{LG}}$.

Liquid droplet concentration in the vapor core is defined as:

$$C = \frac{G_c - G_G}{\frac{G_G}{\rho_G} + \frac{G_c + G_G}{\rho_L}} \tag{1.70}$$

1.5.4 Procedure for Determination of the Location of Dryout

The model development so far has resulted in devising of a sufficient number of equations to determine local values for all mass flow rates distinguished in our study, i.e. G_f, G_c, G_G and δ. Unfortunately, these parameters are interrelated, rendering the equations too complex to be solved analytically. A numerical code is therefore developed to solve these equations using the following procedure:

(1) Equation (1.63) is first used to determine the location of the onset (upstream point) of annular flow. The annular flow region in the micro-channel is then identified as the length from that location. A special marching procedure with iterations at particular location in MathCad14 has been developed enabling that. The initial mass flow rates of the liquid film, liquid droplets, and vapor at the onset of annular flow are evaluated using Eqs. (1.61)–(1.63 ÷ 1.66). The solution is initiated from the node next to the upstream boundary node and proceeds downstream.

   a. Initial film thickness of the annular flow is calculated from relation $\delta_0 = \frac{G_{f0} 2\mu_L}{\tau_w \rho_L}$.
   b. Liquid wall shear at the beginning of annular flow $\tau_{w0} = R_0 \cdot \zeta_0 \cdot \frac{G_f^2}{8 \cdot \rho_L}$
   c. Two-phase flow multiplier at the beginning is determined from the homogeneous flow model and reads $R_0 = 1 + \frac{\rho_L - \rho_G}{\rho_L} \cdot x_0$. In subsequent steps the model due to Muller-Steinhagen and Heck [17] is used with modification Eq. (1.35).
   d. Flow resistance coefficient $\zeta_0 = 0.316 \cdot \mathrm{Re}_f^{-0.25}$.

(2) At each node, an initial guess is made for the liquid film thickness, δ. All other geometrical parameters in the annular flow region can be evaluated using the tentative δ value.

(3) The interfacial shear stress τ_i is evaluated from the formula $\tau_i = f_i \left[\frac{1}{2} \rho_c (u_c - u_L)^2 - \frac{q_i}{2 h_{lv}} (u_c - u_L) \right]$, where the constituting terms are calculated from relations:

a. Friction factor $f_i = 0.005 \cdot \left(1 + 300 \cdot \frac{\delta}{d}\right)$
b. Core velocity $u_c = \frac{G_c}{\rho_c}$
c. Interfacial velocity $u_i = \frac{\tau_w \delta}{2 \mu_L}$
d. Density of vapor core $\rho_c = \frac{1}{\frac{G_G}{G_c} \frac{1}{\rho_G} + \left(1 - \frac{G_G}{G_c}\right) \frac{1}{\rho_L}}$
e. Two other parameters important in the model development are defined as: $x_c = \frac{G_G}{G_c}$ and $x = \frac{G_G}{G}$, where x_c is the effective vapor quality in the vapor core and x-total quality.

(4) All other variables in Eq. (1.60) are evaluated using geometrical parameters determined from step (2) and the mass flow rates from step (3).

(5) The solution seeks to balance both sides of Eq. (1.60). If this criterion is not satisfied, a new value of δ is assumed and steps (2)–(4) repeated until convergence criterion on liquid film mass velocity G_f is reached. Then the values of the primary parameters obtained in the last iteration are adopted for that nodal location.

(6) The code then repeats the same numerical procedure for the next downstream node in pursuit of a balance for G_f at this new node. This procedure is then repeated by marching downstream until the last node in the annular flow region is reached.

1.6 Experimental Facility

An experimental rig has been designed and constructed as a compact, highly integrated mobile unit, Fig. 1.11. It is main part is a closed loop of the working fluid. The rig allows to accomplish both flow boiling and flow condensation experiments. It has been designed as a closed loop of a working medium.

The flow of working fluid is forced by a set of two electrically-powered pumps, connected in series, capable to deliver mass flow rate up to 200 kg/h and gauge pressure up to 8 bars. Gear pumps have been chosen to limit any arising flow pulsations. Adjustment of the mass flow is realized by changing voltage of the pump's power supply or using the by-pass. Working medium is pumped from the main tank through the Danfoss mass flowmeter type MASS D1 3 working with MASS 6000 19″ IP20 interface. Such system, gives about 0.3 % of measurement accuracy.

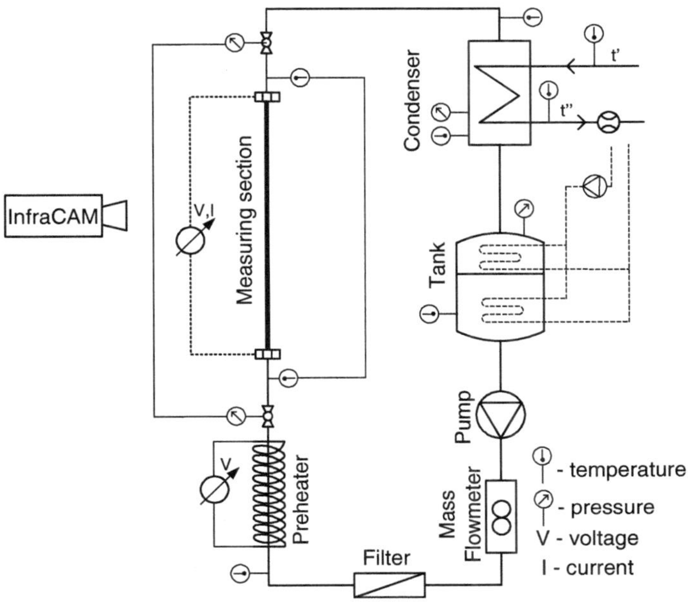

Fig. 1.11 Experimental flow boiling facility

In case of flow boiling experiments the working fluid was directed the pre-heater, where it attained required input parameters. Isobaric pre-heating was realized in the stainless steel tube powered by the low voltage, high current DC power supply. Such arrangement provides the power up to 1.2 kW, corresponding to heat flux of about 170 kW/m^2. The full range of vapour quality x was possible to be obtained at the test section input. Current, voltage, inlet and outlet temperatures and pressure were measured on the pre-heater to determine a corresponding heat flux and quality x from the appropriate heat balance. From the pre-heater the medium went to the test section. In considered experiments silver tubes of 1.15 or 2.3 mm inner diameter, and the length of 38 cm were used. The working medium flowed into the test section with a pre-defined quality x and was heated further to get the expected boiling conditions. Heating in the test section was realized using a low voltage, high current DC power supply and could be adjusted from 0 up to 1 kW of heating power, corresponding to the heat flux up to 364 kW/m^2. Current, voltage, inlet temperature, inlet pressure, outlet temperature and outlet pressure were measured at the test section to determine the corresponding heat flux, sub-cooled liquid temperature, saturation temperature of boiling liquid and pressure drop. The tube wall temperature gradient was measured using a set of ten K-type thermocouples, soldered directly to the wall. All the data were collected automatically with a data acquisition interface. The measuring system used a specially developed in-house software TERMOLAB 06. From the test section the test fluid went to the water cooled condenser and back to the main storage tank. The unit is also equipped with the filter/dryer and an additional pre-heater in the main tank (to be accomplished using hot water).

The strong emphasis was put to the precision of temperature measurements, especially in the light of the fact that in some cases small wall-to-liquid temperature differences were encountered. The K type sheathed thermocouples were used for the measurements of working medium temperature and the K type thermocouple wire was used for the wall temperature measurements. To maximize the accuracy above the standard values quoted by the producer, a special data acquisition interface—TERMOLAB 06—was designed which ensured accuracy of ±0.1 K for the measuring range 0–120 °C and ±0.3 K for the range 120–300 °C. Additionally, each thermocouple was individually calibrated. It was carried out using the stabilized temperature at the accuracy not worse than ±0.05 K. Next, the whole system was tested in 3 constant temperature points, namely 0, 20 and 100 °C. This confirmed the average accuracy of temperature measurement at the level of ±0.1 K for each measuring channel. Maximum deviation between the 10 points on the tube length was also checked in the adiabatic single phase flow conditions and it amounted to ±0.3 K. This value can be admitted as a binding temperature measurement accuracy in the whole experiment.

In Figs. 1.12 and 1.13 presented is modification of the facility to incorporate the possibility of conducting the flow condensation experiments. In that case important is organization of the cooling of the test section. In the present case an approach was developed that a precise heat balance of coolant on the test section was possible. That was made possible by a special design of the cooling section consisting of a series of channels equipped with rotameters on every channel. The details of the test section are shown in Fig. 1.13, whereas the infrared picture of the test section cooling is shown in Fig. 1.14. We can observe a very uniform cooling of every channel, which will lead to precise measurements of heat transfer coefficient and pressure drop.

Input power supplied to the test section (and further—the heat flux) was obtained by multiplying voltage measured on the tube length and current in the tube supplying circuit. Both measurements were carried out using digital multimeters M 890G, whereas the current was measured in the shunt. The deviation of voltage measurement on the tube length was ±0.8 % for the range 20 V AC (resolution 10 mV). The deviation of voltage measurement on the shunt was also ±0.8 % for the range 2 V AC (resolution 1 mV). An error of the shunt resistance measurement and its characteristic determination was assessed as ±1 %. The general error of input power measurements is about ±3 % for the power of about 230 W.

The accuracy of obtained data was assumed by performing an uncertainty analysis based on the method of sequential perturbation (Moffat [71]). This method provides means to estimate the overall uncertainty of data by integrating the uncertainties of individual sources of error into the data base independently, then using a square-root method to calculate the overall uncertainty. Error analysis was performed for every test run of each data set. It was executed automatically by implementing the above procedures in the data reduction spreadsheet. The heat

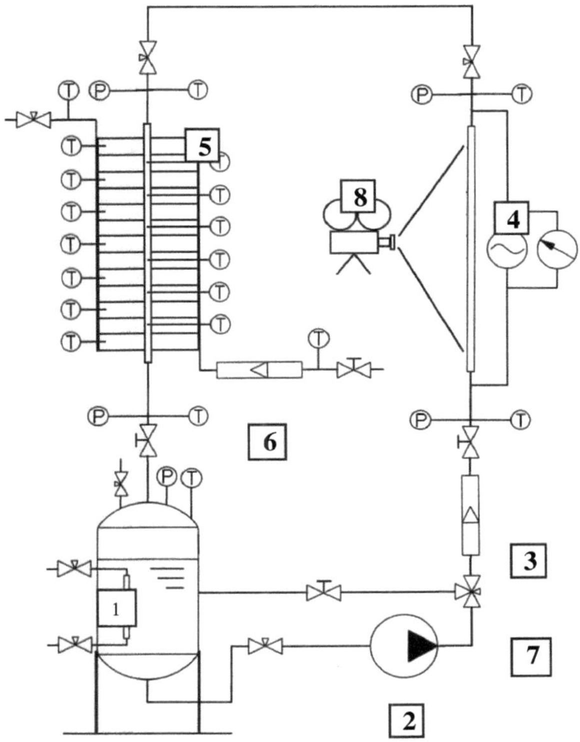

Fig. 1.12 Experimental facility for flow boiling and flow condensation: *1* main tank with pre-heater; *2* circulation pump; *3* mass flow meter; *4* evaporation section; *5* condensation section; *6* rotameter; *7* two-way valve; *8* infrared/ high speed camera

balance was also automatically tested using: temperature difference and heat capacity for the case of single-phase heat exchange and enthalpy difference for the case of flow boiling. The average error of heat transfer coefficient determination not exceeded 5 %. In the flow boiling experiments the R123, R134a, SES36 and ethanol following fluids were studied, whereas in case of condensation: ethanol, R365mfc, HFE7000 and HFE7100, respectively.

The local heat transfer coefficient at each thermocouple point was determined based on the knowledge of local wall temperature T_w, local fluid temperature T_i and the inner wall heat flux to the fluid, q_w:

$$\alpha = \frac{q_w}{T_W - T_i} \tag{1.71}$$

Temperature T_i was deduced from the fluid pressure. This was determined based on the assumption of a linear pressure drop through the test section. T_w was calculated based on the outside surface temperature recorded by the thermocouples, heat flux and the tube wall thermal resistance using a one-dimensional conduction equation. The heat lost to the ambient was included in the calculation. The average error in the heat transfer coefficient was ±7 %. An energy balance based on the heat supplied reduced for losses and the enthalpy change enabled the

Fig. 1.13 Condensation test section. *1* Entry length, *2* Inlet of coolling water, *3* Outlet of cooling water, *4* Inlet of test fluid, *5* Outlet of test fluid

exit thermodynamic quality to be calculated. The total enthalpy change across the test section was calculated based on the flow rate of the test fluid and the pressure and temperature change measured by the differential pressure transducer and thermocouples, respectively, at two ends of the test section. The thermodynamic quality x was determined based on the heat transferred to the fluid, given as:

$$x_i = \frac{h_i - h_l}{h_{lv}} \tag{1.72}$$

where h_l and h_{lv} are the specific enthalpy of saturated liquid and latent heat of vaporisation, respectively. h_i is the local specific enthalpy of the fluid.

The local flow boiling heat transfer coefficients for tested fluids in flow boiling case were obtained for the range: pressure 1 to 6 bar, heat flux 16–170 kW/m^2, mass flux 100–1500 kg/m^2 s, vapour quality 0–0.9. In case of condensation the mass flux range was 200–850 kg/m^2 s, heat flux 45–350 kW/m^2, vapour quality 0–0.9, T_{sat} = 35–90 °C.

Fig. 1.14 Infrared image of test section cooling

1.7 Experimental Validation of Model Predictions

Performance of the postulated model of flow boiling in vertical tubes was presented in a hopefully convincing manner in several journal papers, just to mention Mikielewicz et al. [6], Mikielewicz [7], [57]. In those papers comparisons were made with experimental data from literature, exhibiting a satisfactory performance of the postulated model on a selection of data for conventional size tubes as well as small diameters. A good agreement has also been obtained in case of fluids, for which relation (1.31) was not tested yet, Chiou et al. [60]. A very good report on the superiority of the postulated model for predictions of flow boiling in minitubes was presented by Mahmoud et al. [72]. Presented below is selection of the results of in-house experiments for flow boiling where the peculiar distribution of heat transfer coefficient was found. Additionally presented are comparisons with some empirical correlations from literature. In further parts of paper presented are comparisons with the model of dryout and subsequently flow condensation.

1.7.1 M-Shape Distribution of Heat Transfer Coefficient in Flow Boiling in Minichannels

In flow boiling in conventional size channels there is observed a maximum in heat transfer coefficient distribution in function of quality which is found at about x ≅ 0.8. That location corresponds to the existence of annular flow structure in the tube. In case of small diameter channels that maximum moves towards smaller qualities. In Fig. 1.15 presented is a qualitative distribution of heat transfer coefficient which has been observed during some of the experiments in literature, see for example Bar-Cohen and Rahim [73].

In case of channel diameters equal to 2.3 mm that maximum was found for x ≅ 0.4 and in case of even smaller diameters at x ≅ 0.1, Zrooga [74]. That has also been confirmed by recent research [5], who confirm that the bubbly flow is found in small diameter channels only up to qualities not exceeding x ≅ 0.1. In studies of flows in minichannels there can also exist an additional maximum of heat transfer coefficient in its distribution in function of quality. The only known study thus far, where such situation was noticed is the paper by Bar-Cohen and Rahim [73]. Similar findings were, however, also found in the recent study by Zrooga [74] and Klugmann [75].

The mechanism of development of M-shape distribution can be explained in the following way. Values of heat transfer coefficient gradually increase from the values which are obtained at small subcoolings to the local maximum close to x = 0. Then, the decrease is observed followed by a plateau and subsequently by a slight increase of heat transfer coefficient. At values of quality about 60 % the second maximum is observed. After that another decrease of heat transfer coefficient is found. The smallest values are reached when x approaches value of unity, which is generally related to the dryout conditions. The M-shape distribution of heat transfer coefficient seems to be a specific phenomenon devoted merely to two-phase flows in minichannels. Initiation of boiling on the wall and related to it acceleration of saturated liquid flow induces significant rise of heat transfer coefficient, significantly above the level of single phase values. That may explain the first steep rise of the heat transfer coefficient distribution. Subsequent transition to the slug/plug flow structure renders gradual decrease of heat transfer coefficient, as the vapour slugs form conditions for local evaporation of thin liquid film, separating the liquid and the wall, and even to development of dry patches, which impair heat transfer. The transition to annular flow structure again results in increase of heat transfer coefficient, as a result of development of evaporating thin liquid film on the wall. The evaporating film thinners and splits into rivulets causing local dryout conditions. It is very rare, if possible, to find in literature distributions of heat transfer coefficient in a whole range of quality variation. Usually the experimental data refer to a range of quality and enable to reflect only a part of the M-shaped curve.

A detailed analysis of experimental data enables to conclude that different sets of data by various authors, which have been published on that topic describe

Fig. 1.15 Schematic of heat transfer distribution in conventional size channel and minichannels

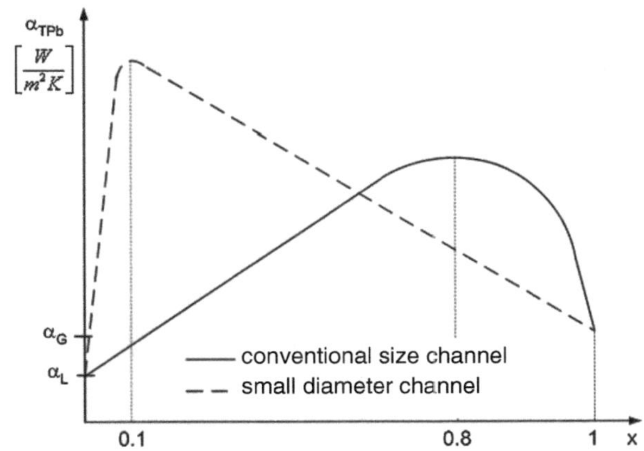

exactly different parts of the M-shape distribution of heat transfer coefficient. That is the reason why in some cases authors are claiming the reduction of heat transfer coefficient with quality, for example Kandlikar [1], and sometimes the increasing trend, Shiferaw et al. [76].

An abundant amount of experimental evidence has been analysed by Thome [77]. He concluded that for small qualities in the flow when the bubbly flow structure is dominant the transport of heat takes place merely by generated bubbles. In case of greater qualities, corresponding to the long slugs separated by liquid then heat is transported by a combined convective-conductive mechanism in the thin layer of liquid film surrounding the slug. In such case values of heat transfer coefficient are very strongly dependent on the downstream translational velocity of vapour slugs. In case of high values of quality, characteristic to annular flow structure, the mechanism of heat transfer is dependent on convection and conduction related to film evaporation from the wall (Jacobi et al. [78]). The second maximum, however, has not been detected in those studies. The experimental challenge is that it is very difficult, if at all possible, to measure the values of heat transfer coefficient for very small and very large values of quality.

Analysis of Figs. 1.16 and 1.17 shows that the heat transfer coefficient depends on the value of heat flux, at constant value of mass velocity. In Fig. 1.16 presented are results for ethanol, whereas in Fig. 1.17 for R123. Experimental data show some discrepancy but distributions of heat transfer coefficient conform to some trends. The experimental discrepancy is related to the fact that there could be small pressure pulsations in operation of the pumps. Also a strongly non-equilibrium character of considered processes could influence the readings. There ought to be noticed, however, the presence of two maxima of heat transfer coefficient, one for $x \approx 0.1 \div 0.3$ and the one for $x \approx 0.6 \div 0.7$, which is consistent with findings of Klugmann [75] for R123.

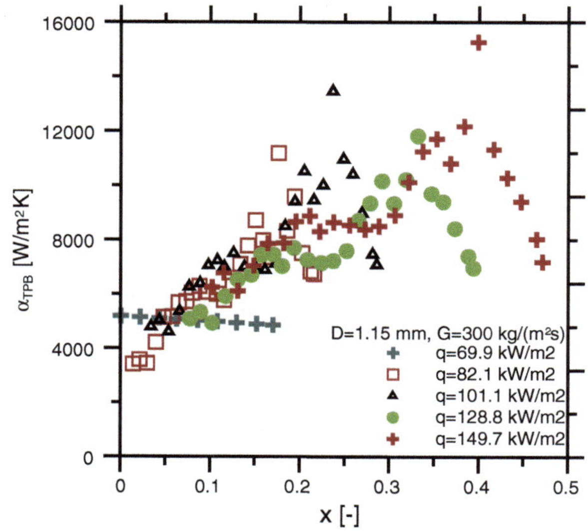

Fig. 1.16 Influence of heat flux on local heat transfer coefficient in function of quality, ethanol, d = 1.15 mm, G = 300 kg/(m² s), [74]

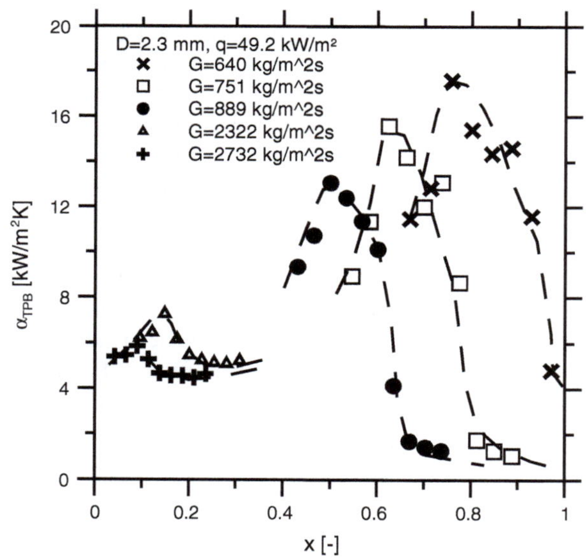

Fig. 1.17 Influence of heat flux on local heat transfer coefficient in function of quality, R123, d = 2.3 mm, q = 49246 W/m², [75]

The maximum corresponding to smaller qualities shifts towards higher values of vapour content with increasing heat flux. There was a significant difficulty in obtaining the systematic data for higher qualities.

It can be noticed that the presented picture of the influence of heat flux at a set value of mass velocity is not monotonic in function of quality. For small qualities ($x < 0.4$) it can be seen that with the increase of heat flux the heat transfer coefficient increases. On the other hand for greater values of quality the reverse trend is observed, which points at the existence of the second local maximum, which is in line with the investigations of Bar-Cohen and Rahim [73]. After the second maximum all distributions indicate that the heat transfer coefficient decreases with increase of quality.

1.7.2 Flow Boiling in Small Diameter Tubes and Minichannels

Comparisons with data from literature were presented also in mentioned above papers, namely Mikielewicz et al. [6], Mikielewicz [7], [57]. In the present section some comparisons will be shown with the data obtained on the facility presented earlier for laminar (Zrooga [74]) and turbulent (Klugmann [75]) flow boiling flows of ethanol and R123 respectively.

In Fig. 1.18 presented are the results of measurements of flow boiling of ethanol and compared with the predictions obtained using the model due to Mikielewicz [6, 57], see relation (1.31). Over 80 % of data fall into the error band of ±30 and 95 % within ±50 %. That error is evenly distributed for all qualities, Fig. 1.19. In case of small qualities, the consistency between measured data and calculations performed using the model (1.31) is very good. The consistency slightly decreases for greater qualities.

In literature there is a large number of empirical correlations for description of saturated flow boiling. Their best performance is obtained in case when the correlations are applied to their developers data. Usually their application to other fluids than originally tested is a failure. For consideration here three correlations have been taken, which are specifically developed for flow boiling in minichannels. These are correlations due to Steinke and Kandlikar [40] (1.18a, b), Lazarek and Black [39] (1.17), and Owhaib [79]. The experimental results have also been also compared against the model described by (1.31). Comparisons are shown in Figs. 1.20 and 1.21.

The superiority of the model described by Eq. (1.31) can be seen from the attached figures. It can be said that only in selected cases other correlations perform to a satisfactory extent. The correlation (1.31) can describe the increasing trend of heat transfer coefficient as well as the decreasing one.

1.7.3 Comparisons of Flow Condensation Predictions with Experimental Data

Presented below is comparison of selected correlations for calculations of flow condensation with the model presented in the first part of the paper, namely relation (1.31). As the proposed model was thoroughly tested for the conditions of flow

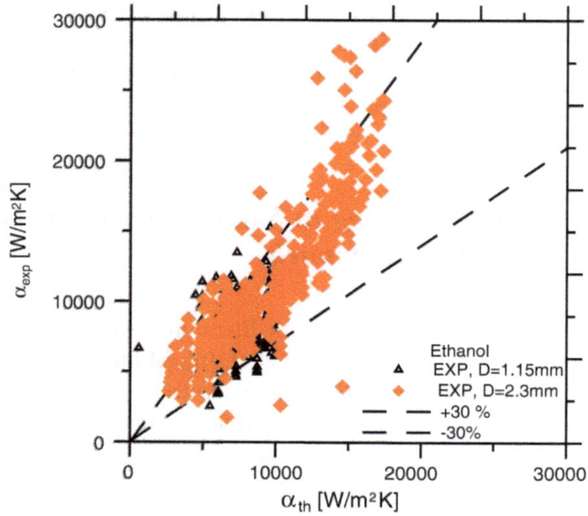

Fig. 1.18 Experimental heat transfer coefficient vz. heat transfer coefficient calculated by using (1.31), ethanol, d = 1.15 and 2.3 mm

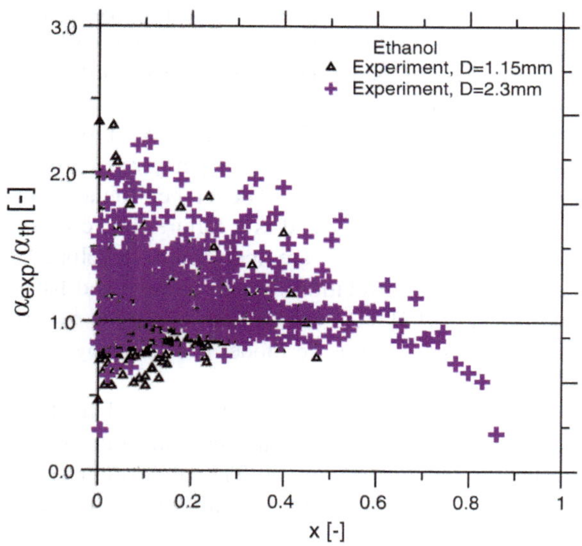

Fig. 1.19 Ratio of experimental heat transfer coefficient to theoretical one in function of quality, ethanol, d = 1.15 and 2.3 mm

boiling showing satisfactory performance, that encouraged author to pursuit further the extension of the model to flow condensation conditions. Obviously in that case the full form of the flow boiling correlation (1.31) cannot be used, as in that case the bubble generation is not present. Comparisons have been carried out for different fluids for data collected from literature. The results of calculations of heat transfer coefficient against quality are presented in Figs. 1.22, 1.23, 1.24 and 1.25.

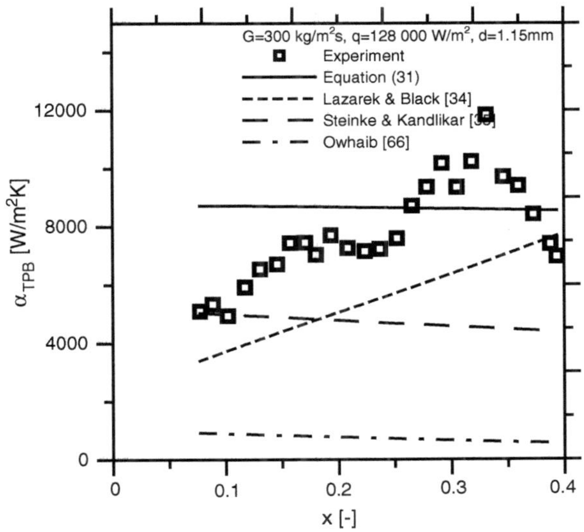

Fig. 1.20 Comparison of different correlation with data in function of quality, ethanol, $G = 300\ kg/m^2\ s$, $q = 128\ 000\ W/m^2$, $D = 1.15\ mm$

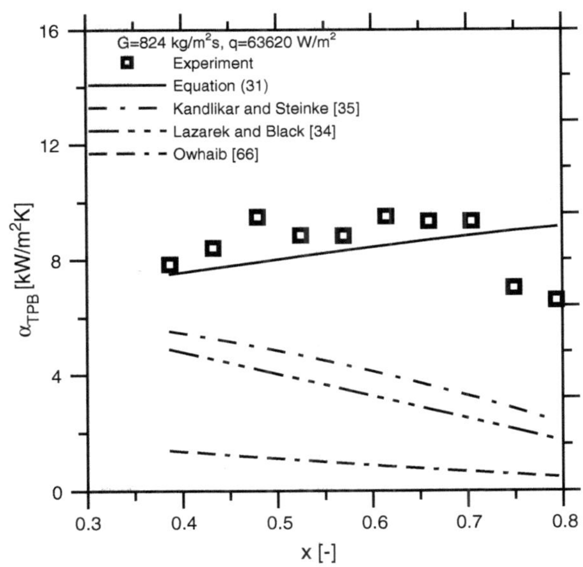

Fig. 1.21 Comparison of different correlations with data in function of quality, R123, $G = 824\ kg/m^2\ s$, $q = 63620\ W/m^2$, $d = 2.3\ mm$

It can be seen that Eq. (1.31) describes reasonably well the heat transfer coefficients during the flow condensation. First of all, the model of heat transfer coefficient with non-adiabatic correction performs qualitatively correctly, confirming the fact that the heat transfer is a function of the two-phase pressure drop. The biggest discrepancy is found in case of higher qualities. That could probably

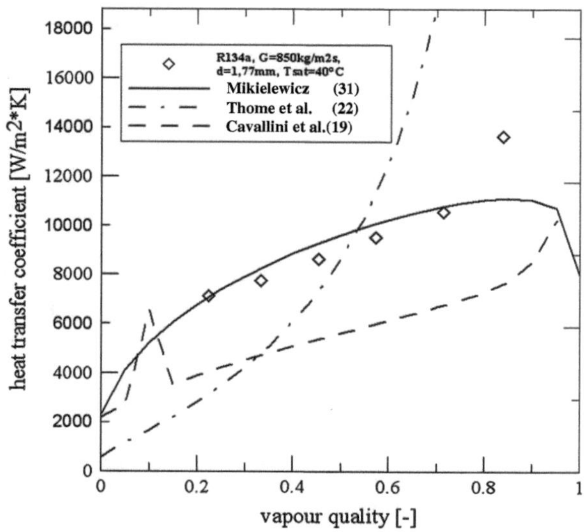

Fig. 1.22 Comparison of heat transfer coefficient for R134a in relation to experiment [66]; d = 1.77 mm, G = 850 kg/m^2 s, Tsat = 40 °C

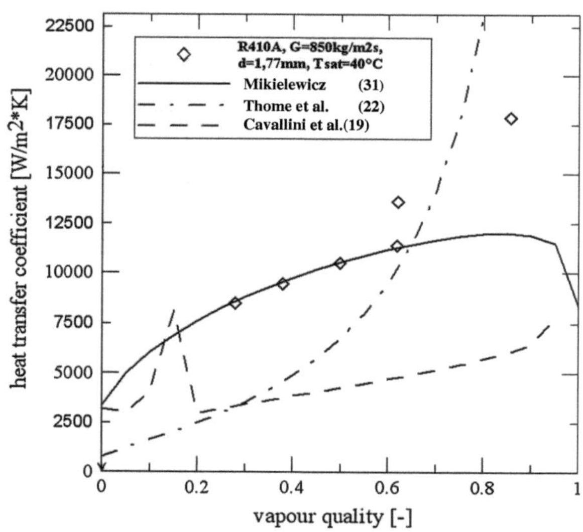

Fig. 1.23 Comparison of heat transfer coefficient for R410a in relation to experiment [66]; d = 1.77 mm, G = 850 kg/m^2 s, Tsat = 40 °C

be improved if better models for the two-phase flow multiplier and hence pressure drop are developed. The exact behaviour of the blowing parameter should also attract further and more comprehensive attention as the phenomenon of condensation may require more detailed description of velocity changes at the interface.

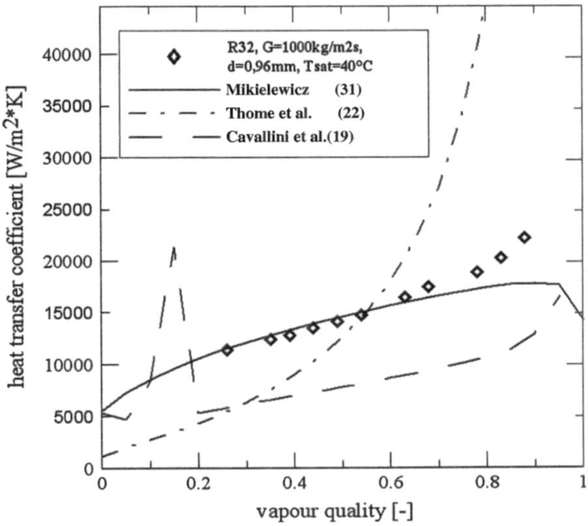

Fig. 1.24 Comparison of heat transfer coefficient for R32 in relation to experimental data [80]; d = 0,96 mm, G = 1000 kg/m^2 s, T_{sat} = 40 °C

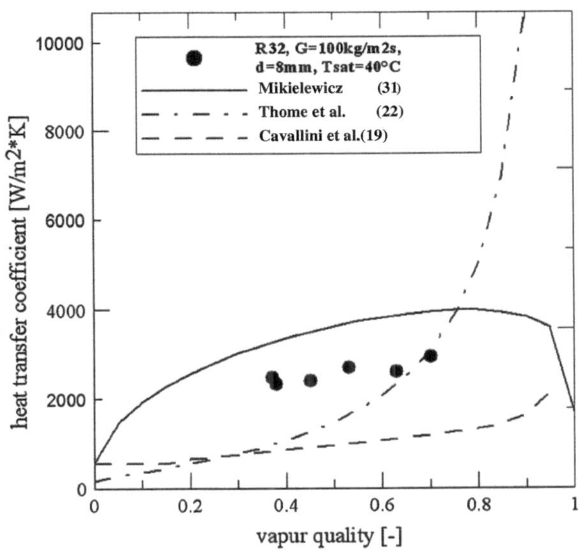

Fig. 1.25 Comparison of heat transfer coefficient for R32 in relation to experimental data [81]; d = 8 mm, G = 100 kg/m^2 s, T_{sat} = 40 °C

1.7.4 Comparison of Dryout Results with Experimental Data

Verification of models was accomplished for tubes having diameters of d = 1.15 and 2.3 mm, with ethanol as a test fluid, Zrooga [74]. Obtained experimentally results were compared with predictions of the five equation model, as well as the three equation model [70], where only the mass balance equations were considered.

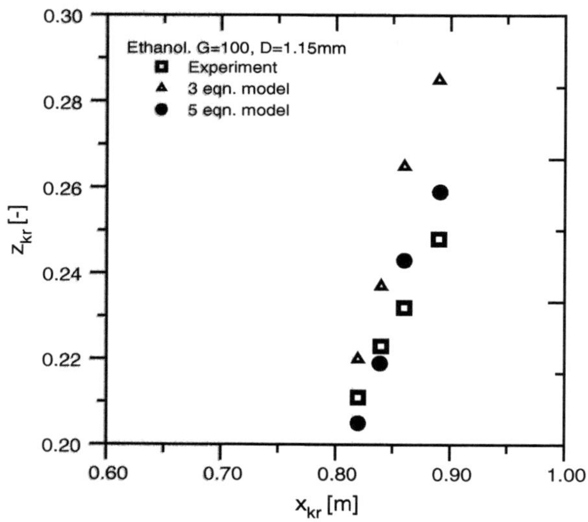

Fig. 1.26 Results of calculations obtained using the 5 equation model with experiments for ethanol in tubes. d = 1.15 mm, G = 100 kg/m^{2}s

In Figs. 1.26 and 1.27 presented is a comparison of a five equation model predictions in comparison to experimental data for ethanol for the tube diameter d = 1.15 and 2.3 mm.

In all presented cases we can observe a superiority of the postulated 5 equation model over the 3 equation model in relation to obtained experimental data. Better consistency is obtained in case of the smaller tube size, i.e. d = 1.15 mm. In that case the superiority of a new model is apparent. In case of the tube with diameter d = 2.3 mm the new model still returns better consistency, but the discrepancy is bigger.

The reason for a better agreement is that the momentum equation provides us with a more accurate value of the liquid film thickness, which subsequently enables more accurate determination of the dryout location. Calculations of the dryout location are very sensitive to the selection of the deposition correlation. In the course of the study several models have been tested, but only the correlation due to Sugawara (Okawa et al. [82]) was found to work best with the model.

Another issue is that obtained in the experiment flow conditions relate to laminar flows, whereas the correlations used for deposition are mostly corresponding to turbulent flow conditions. There is a lack of appropriate deposition and entrainment correlations for laminar flows.

Therefore it can be said that further intense research in that area is necessary. Inclusion of momentum equations for liquid and vapour leads to a better determination of the liquid film thickness, which is crucial in the correct determination of location of dryout. The new model has been elaborated, however it uses the correlations, which have not been specifically tailored to laminar flows and that is the reason for some discrepancy.

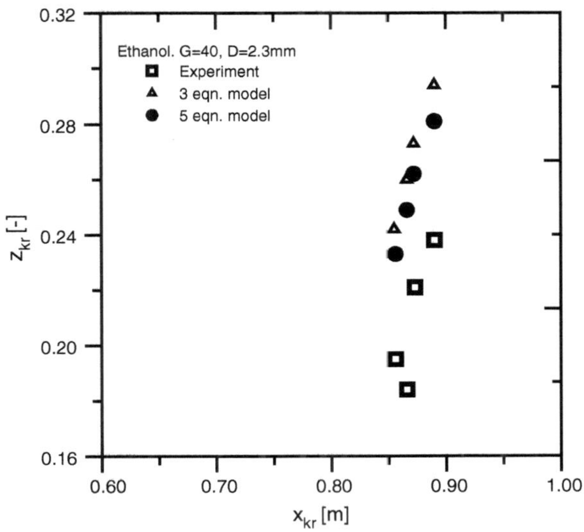

Fig. 1.27 Results of calculations obtained using the 5 equation model with experiments for ethanol in tubes. d = 2.3 mm, G = 40 kg/m^2 s

It is obvious that in future the agreement of the presented model of dryout predictions and experimental data can be improved if new and better correlations will be developed for deposition of respective phases. Questionable still seems the issue of entrainment in minichannels, which for the time being is neglected from the considerations in laminar flows. There is no clear indication from the existing experimental data, whether the entrainment is important in laminar flows or can be neglected, as suggested by for example Qu and Mudawar [68].

1.8 Summary

In the book some issues of thermal-hydraulics modeling in minichannels with change of phase were presented. These were:

1. common modeling of flow boiling and flow condensation using the same correlation equation,
2. peculiar effect of M-shape distribution of heat transfer coefficient in flow boiling in minichannels,
3. theoretical modeling of dryout phenomenon in minichannels.

The model of flow boiling and flow condensation presented in the paper is a general model, which can be advised for use in majority of engineering applications. Comparisons with different experimental data show its robustness and satisfactory accuracy. The model has been originally developed for flow boiling. Contrary to other methods of calculating heat transfer coefficients in flow boiling, it has been developed on the theoretical basis, as a result of consideration of energy dissipation in the flow, where boiling occurs. That led, first of all, to devising a value of exponent n in Eq. (1.16), most often used for reduction of experimental data. As results from the presented model the value of exponent n = 2. The postulated model also incorporates another term, which proves to be very important in case of minichannels and microchannels. That is the inclusion of a two-phase flow multiplier, which models the convective flow without bubble nucleation. The selection of the two-phase flow multiplier is very important for calculations. In case of refrigerants, the most effective is a model due to Müller-Steinhagen and Heck [17]. In case of other fluids the latter model may not show to provide good accuracy of calculations. The more accurate model for two-phase flow multiplier will be selected the more accurate results will be obtained. In order to get the most accurate results of calculations we must use the models of pressure drop developed for specific flow structures. That proves that the approach to model heat transfer in flow boiling starting from precise evaluation of pressure drop is very promising, but that is also consistent with the general structure of presented here model, with the two-phase flow multiplier being an important term in it. What can be advised for perspective research into flow boiling is that the measurements of pressure drop and quantities constituting the heat transfer coefficient should take place simultaneously, in order to obtain the most accurate information.

Presented here model is applicable both to conventional channels and small diameter channels. Overall, it has been tested for the channel diameters greater than 1 mm. In case of smaller channel dimensions the presented model tends to overpredict the experimental data, however still presents robustness. In case of minichannels smaller than 1 mm it may be recommended to drop the entire nucleate boiling term in Eq. (1.16). In effect will the dependence of two-phase flow heat transfer coefficient will be merely on the two-phase flow multiplier, which would serve as a useful information to other researchers, as to how continue their investigations. In such case crucial is appropriate selection of two-phase flow multiplier.

In author's opinion, the presented model can be suggested for a wider use amongst engineers, but further validation with experimental data would add value to its robustness.

In this chapter presented is also a comparison of predictions of condensation inside channels with the correlation developed for flow boiling and examined for that purpose. The comparisons were made with correlations developed for the annular flow structure, where flow boiling and flow condensation can be regarded as symmetrical phenomena. The comparisons were also carried out with some experimental data from literature showing satisfactory agreement.

Additionally studied were the non-adiabatic effects in the description of the model of heat transfer coefficient (1.31). The blowing parameter has been incorporated into modeling of annular flow heat transfer as well as another correction was added for heat flux dependence in other than annular flows. The modification leads in case of condensation to biggest effects for small qualities, confirming the fact that non-adiabatic effects for higher qualities are of smaller importance. Also in case of flow boiling the biggest modifications to heat transfer coefficient occur for smaller qualities (smaller than 0.4). That issue will require more attention, as the approach presented in the work is too simplistic. The distribution of the blowing parameter with the change of film thickness is required to improve further the accuracy of predictions of the flow boiling/flow condensation model.

Presented has also been an original five equation model of dryout in mini-channels. The consideration of momentum equations for the liquid in the film as well as liquid and gas in the core forms a step forward in modeling of dryout in small diameter tubes. The developed five equation model of dryout [74] has been compared with acquired experimental data for considered channels showing a superiority in relation to the three-equation model developed earlier by Glinski [70]. The developed model enables calculation of liquid film thickness and such updated value is used in calculations of mass balance in liquid film and the core.

In future developments of the presented model of flow boiling required is incorporation of the correct model of subcooling and post-dryout heat transfer. That issue will be scrutinized in the future.

Acknowledgments The work presented in the paper was partially funded from the Polish Ministry for Science and Education research project No. N512 459036 in years 2009–2012.

References

1. S.G. Kandlikar, Fundamental issues related to flow boiling in minichannels and microchannels, in *Proceedings of Experimental Heat Transfer, Fluid Mechanics and Thermodynamics*, Thessaloniki, pp. 129–146 (2001)
2. P. Kew, K. Cornwell, Correlations for the prediction of boiling heat transfer in small diameter channels. Appl. Therm. Eng. **17**(8–10), 705–715 (1997)
3. L. Cheng, G. Ribatski, J. Moreno-Quibén, J.R. Thome, New prediction methods for CO_2 evaporation inside tubes: part I—a two-phase flow pattern map and a flow pattern based phenomenological model for two-phase flow frictional pressure drops. Int. J. Heat Mass Transfer **51**, 111–124 (2008)
4. L. Cheng, G. Ribatski, J. Moreno-Quibén, J.R. Thome, New prediction methods for CO_2 evaporation inside tubes: part II—a two-phase flow pattern map and a flow pattern based phenomenological model for two-phase flow frictional pressure drops. Int. J. Heat Mass Transf. **51**, 125–135 (2008)
5. J.R. Thome, L. Consolini, Mechanisms of boiling in micro-channels: critical assessment, in *Proceedings of 5th International Conference on Transport Phenomena in Multiphase Systems*, 30 June 30–3 July 2008, Bialystok, Poland (2008)
6. D. Mikielewicz, J. Mikielewicz, J. Tesmar, Improved semi-empirical method for determination of heat transfer coefficient in flow boiling in conventional and small diameter tubes. Int. J. Heat Mass Transf. **50**, 3949–3956 (2007)
7. D. Mikielewicz, A new method for determination of flow boiling heat transfer coefficient in conventional diameter channels and minichannels. Heat Transf. Eng. **31**(4), 276–287 (2010)
8. D. Baker, Simultaneous flow of oil and gas. Oil Gas J **53**, 183–195 (1954)
9. Y. Taitel, D. Barnea, A.E. Dukler, Modeling flow pattern transitions for steady upward gas–liquid flow in vertical tubes. AIChE J. **26**(3), 345–354 (1980)

10. G.F. Hewitt, D.N. Roberts, Studies of two-phase flow patterns by simultaneous X-ray and flash photography, AERE-M 2159, London (1969)
11. T. Oshinowo, M.E. Charles, Vertical two-phase flow. Part I. Flow pattern correlations. Can. J. Chem. Eng. **52**, 25–35 (1974)
12. K.A. Triplett, S.M. Ghiaasiaan, S.I. Abdel-Khalik, D.L. Sadowski, Gas–liquid two-phase flow in microchannels. Part I: two-phase flow patterns. Int. J. Multiphase Flow **25**, 377–394 (1999)
13. P.M.-Y. Chung, M. Kawaji, The effect of channel diameter on adiabatic two-phase flow characteristics in microchannels. Int. J. Multiphase Flow **30**, 735–761 (2004)
14. S.M. Zivi, Estimation of steady state steam void fraction by means of the principle of minimum entropy production. Trans. ASME (J. Heat Transf.) **86**, 247–252 (1964)
15. R.W. Lockhart, R.C. Martinelli, Proposed correlation of data for isothermal two-phase two-component flow in pipes. Chem. Eng. Prog. **45**, 39–45 (1949)
16. D. Chisholm, Pressure gradients due to friction during the flow of evaporating two-phase mixtures in smooth tubes and channels. Int. J. Heat Mass Transf. **16**, 347–358 (1973)
17. R. Muller-Steinhagen, K. Heck, A simple friction pressure drop correlation for two-phase flow in pipes. Chem. Eng. Prog. **20**, 297–308 (1986)
18. L. Friedel, Improved friction pressure drop correlations for horizontal and vertical two-phase pipe flow, in *European Two-Phase Flow Group Meeting*, Paper E2, Ispra, Italy (1979)
19. K.M. Mishima, T. Hibiki, Some characteristics of air–water two-phase flow in small diameter vertical tubes. Int. J. Multiphase Flow **22**(4), 703–712 (1996)
20. T.N. Tran, M.-C. Chyu, M.W. Wambsganss, D.M. France, Two-phase pressure drop of refrigerants during flow boiling in small channels: an experimental investigation and correlation development. Int. J. Multiphase Flow **26**, 1739–1754 (2000)
21. M. Zhang, R.L. Webb, Correlation of two-phase friction for refrigerants in small-diameter tubes. Exp. Therm. Fluid Sci. **25**, 131–139 (2001)
22. J.R. Thome, Boiling of new refrigerants: a state-of-the-art review". Int. J. Refrig. **19**, 435–457 (1996)
23. A.E. Bergles, V.J.H. Lienhard, G.E. Kendall, P. Griffith, Boiling and evaporation in small diameter channels. Heat Transf. Eng. **24**(1), 18–40 (2003)
24. J.R. Thome, Boiling in microchannels: a review of experiment and theory. Int. J. Heat Fluid Flow **25**, 128–139 (2004)
25. C.E. Dengler, J.N. Addoms, Heat transfer mechanism for vaporisation of water in vertical tube. Chem. Eng. Prog. Symp. Ser. **52**(18), 95–103 (1956)
26. S.A. Guerrieri, R.D. Talty, A study of heat transfer to organic liquids in single tube natural circulation vertical tube boilers. Chem. Eng. Prog. Symp. Ser. **52**(18), 69–77 (1956)
27. V.E. Schrock, L.M. Grossman, Forced convection boiling studies. Institute of Engineering Research, University of California, Report 73308-UCX-2182 (1959)
28. J.G. Collier, D.J. Pulling, Heat transfer to two-phase gas-liquid systems", Report AERE-R3908 (1962)
29. M.M. Shah, Chart correlation for saturated boiling heat transfer: Equations and further study. ASHRAE Trans. **88**, I, 185–196 (1982)
30. S.G. Kandlikar, A general correlation for saturated two-phase flow boiling heat transfer inside horizontal and vertical tubes. J. Heat Transf. **112**, 219–228 (1989)
31. K.E. Gungor, R.H.S. Winterton, A general correlation for flow boiling in tubes and annuli". Int. J. Heat Mass Transf. **29**, 351–358 (1986)
32. W.M. Rohsenow, A method of correlating heat transfer data for surface boiling of liquids. Trans. ASME **74**, 969 (1952)
33. J.C. Chen, Correlation for boiling heat-transfer to saturated fluids in convective flow. Ind. Chem. Eng. Proc. Des. Dev. **5**(3), 322–339 (1996)
34. S.S. Kutateładze, Boiling heat transfer. Int. J. Heat Mass Transfer **4**, 31–45 (1961)
35. D. Steiner, J. Taborek, Flow boiling heat transfer in vertical tubes correlated by asymptotic model. Heat Transf. Eng. **23**(2), 43–68 (1992)

36. H.K. Forster, N. Zuber, Dynamics of vapour bubbles and boiling heat-transfer. AIChE J. **1**, 531–535 (1955)
37. M.G. Cooper, Saturation nucleate pool boiling: a simple correlation. Int. Chem. Eng. Symp. **86**, 785–793 (1984)
38. J. Mikielewicz, Semi-empirical method of determining the heat transfer coefficient for subcooled saturated boiling in a channel. Int. J. Heat Transf. **17**, 1129–1134 (1973)
39. G.M. Lazarek, S.H. Black, Evaporative heat transfer, pressure drop and critical heat flux in a small vertical tube with R-113. Int. J. Heat Mass Transf. **25**(7), 945–960 (1982)
40. M. Steinke, S. Kandlikar, An experimental investigation of flow boiling characteristics of water in parallel microchannels. J. Heat Transf. **126**(4), 518–526 (2004)
41. N. Kattan, J.R. Thome, D. Favrat, Flow boiling in horizontal tubes: part 3—development of a new heat transfer model based on flow pattern, Trans. ASME, **120**, 156–165 (1998)
42. L. Wojtan, T. Ursenbacher, J.R. Thome, Development of a new heat transfer model for stratified-wavy, dryout and mist flow regimes. Int. J. Heat Mass Transf. **48**, 2970–2985 (2005)
43. R.T. Lahey, D.A. Drew, An analysis of two-phase flow and heat transfer using a multidimensional, multi-field, two-fluid computational fluid dynamics (CFD) model, in *Japan/US Seminar on Two-Phase Flow Dynamics*, 5–8 June 2000, Santa Barbara, California, (2000)
44. M. Podowski, Understanding multiphase flow and heat transfer: perception, reality, future needs. Arch. Thermodyn. **26**(3), 3–20 (2005)
45. A. Cavallini, G. Censi, D. Del Col, L. Doretti, G.A. Longo, L. Rossetto, Condensation of halogenated refrigerants inside Smooth tubes. HVAC Res. **8**(4), 429–451 (2002)
46. J. El Hajal, J.R. Thome, A. Cavallini, Condensation in horizontal tubes, part 1: two-phase flow pattern map. Int. J. Heat Mass Transf. **46**(18), 3349–3363 (2003)
47. J.R. Thome, J. El Hajal, A. Cavallini, Condensation in horizontal tubes, part 2: new heat transfer model based on flow regimes. Int. J. Heat Mass Transf. **46**(18), 3365–3387 (2003)
48. S. Garimella, Condensation flow mechanisms in microchannels: basis for pressure drop and heat transfer models. Heat Transf. Eng. **25**(3), 104–116 (2004)
49. W.W. Akers, H.A. Deans, O.K. Crosser, Condensation heat transfer within horizontal tubes. Chem. Eng. Prog. Symp. Ser. **55**(29), 171–176 (1959)
50. M.M. Shah, A general correlation for heat transfer during film condensation inside pipes. Int. J. Heat Mass Transf. **22**(4), 547–556 (1979)
51. D.P. Traviss, W.M. Rohsenow, Flow regimes in horizontal two-phase flow with condensation. ASHRAE Trans. **79**(Part 2), 31–39 (1973)
52. M.K. Dobson, J.C. Chato, Condensation in smooth horizontal tubes. J. Heat Transf. Trans. ASME **120**(1), 193–213 (1998)
53. A. Cavallini, G. Censi, D. Del Col, L. Doretti, G.A. Longo, L. Rossetto, C. Zilio, Condensation inside and outside smooth and enhanced tubes—a review of recent research". Int. J. Heat Mass Transf. **26**, 373–392 (2002)
54. G. Breber, J.W. Palen, J. Taborek, Prediction of horizontal tubeside condensation of pure components using flow regime criteria. J. Heat Transf. Trans. ASME **102**(3), 471–476 (1980)
55. P.G. Kosky, F.W. Staub, Local condensing heat transfer coefficients in the annular flow regime. AIChE J. **17**(5), 1037–1043 (1971)
56. D. Mikielewicz, J. Mikielewicz, A common method for calculation of flow boiling and flow condensation heat transfer coefficients in minichannels with account of non-adiabatic effects. Heat Transf. Eng. **32**(10), 1173–1181 (2011)
57. E.L. Ananiev, On the mechanism of heat transfer in nucleate boiling flow of water in a tube against the Reynolds analogy, *Convective Heat Transfer in Single and Two-Phase Flows* ed. by V.M. Borishansky (Energia, Leningrad, 1964)
58. M.B. Ould Didi, N. Kattan, J.R. Thome, Prediction of two-phase pressure gradients of refrigerants in horizontal tubes. Int. J. Refrig. **25**, 935–947 (2002)
59. L. Sun, K. Mishima, Evaluation analysis of prediction methods for two-phase flow pressure drop in mini-channels. Int. J. Multiphase Flows **35**, 47–54 (2009)

60. C.B. Chiou, D.C. Lu, C.Y. Liao, Y.Y. Su, Experimental study of forced convective boiling for non-azeotropic refrigerant mixtures R-22/R-124 in horizontal smooth tube. Appl. Therm. Eng. **29**, 1864–1871 (2009)
61. J. Mikielewicz, Influence of phase changes on shear stresses at the interfaces. Trans. Inst. Fluid Flow Mach. **76**, 31–39 (1978). (in Polish)
62. S.S. Kutateładze, A.I. Leontiev, *Turbulent boundary layers in compressible gases* (Academic Press, NY, 1964)
63. G.B. Wallis, One dimensional two-phase flow (McGraw-Hill, New York, 1969)
64. L. Bai, G. Lin, H. Zhang, D. Wen, Mathematical modelling of steady-state operation of a loop heat pipe. Appl. Therm. Eng. **29**, 2643–2654 (2009)
65. T. Bohdal, H. Charun, M. Sikora, Comparative investigations of the condensation of R134a and R404A refrigerants in pipe minichannels. Int. J. Heat Mass Transf. **54**, 1963–1974 (2011)
66. A. Cavallini, G. Censi, D. Del Col, L. Doretti, G.A. Longo, L. Rossetto, Experimental investigation on condensation heat transfer and pressure drop of new HFC refrigerants (R134a, R125, R32, R410A, R236ea) in a horizontal smooth tube. Int. J. Heat Mass Transf. (2007)
67. J.G. Collier, J.R. Thome, *Convection Boiling and Condensation*, 3rd edn. (McGraw-Hill, New York, 1994)
68. W. Qu, I. Mudawar, Flow boiling heat transfer in two-phase micro-channel heat sink—II. Annular two-phase flow model. Int. J. Heat Mass Transf. **46**, 2773–2784 (2003)
69. R. Revellin, P. Haberschill, J. Bonjour, J.R. Thome, Conditions of liquid film dryout during saturated flow boiling in microchannels. Chem. Eng. Sci. **63**, 5795–5801 (2008)
70. M. Gliński, Study of critical heat flux in small diameter channels. PhD Thesis. Gdansk University of Technology (2010)
71. R.J. Moffat, Describing the uncertainties in experimental results". Exp. Therm. Fluid Sci. **1**, 3–17 (1988)
72. M.M. Mahmoud, D.B. Kenning, T. Karayiannis, Surface effects and evaluation of prediction methods in boiling flow of R134a in microtubes, in *48th European Two-Phase Flow Group Meeting*, 28 June–1 July 2010, Brunel University, London (2010)
73. A. Bar-Cohen, E. Rahim, Modelling and prediction of two-phase microgap channel heat transfer characteristics. Heat Transf. Eng. **38**(8), 601–625 (2010)
74. A.-B.R. Zrooga, Experimental and theoretical study of flow boiling and dryout phenomenon of ethanol in vertical minitubes. PhD Thesis, Gdansk University of Technology (2010)
75. M. Klugmann, Heat transfer intensification in flow boiling in small diameter channels. PhD Thesis, Gdansk University of Technology (in Polish) (2009)
76. D. Shiferaw, X. Huo, T.G. Karayiannis, D.B.R. Kenning, Examination of heat transfer correlations and a model for flow boiling of R134a in small diameter tubes. Int. J. Heat Mass Transf. **50**, 5177–5193 (2007)
77. J.R. Thome, Chapter I, *Wolverine Engineering Databook III* (2007). www.wlv.com/products
78. A.M. Jacobi, J.R. Thome, Heat transfer model for evaporation of elongated bubble flows in microchannels. J. Heat Transf. **124**, 1131–1136 (2002)
79. W. Owhaib, Experimental heat transfer, pressure drop and flow visualization of R134a in vertical mini/micro tubes. PhD Thesis, KTH, Stockholm (2007)
80. H.-K. Oh, C.-H. Son, Condensation heat transfer characteristics of R-22, R-134a and R-410A in a single circular microtube, Exp. Therm. Fluid Sci. **35**, 706–716 (2011)
81. M. Matkovic, A. Cavallini, D. Del Col, L. Rossetto, Experimental study on condensation heat transfer inside a single circular minichannel. Int. J. Heat Mass Transf. **52**, 2311–2323 (2009)
82. T. Okawa, A. Kotani, I. Kataoka, M. Naitoh, Prediction of the critical heat flux in annular regime in various vertical channels. Nucl. Eng. Des. **229**, 223–236 (2004)

Chapter 2
Flow Characteristics and Void Fraction Prediction in Large Diameter Pipes

Xiuzhong Shen, Joshua P. Schlegel, Shaowen Chen, Somboon Rassame, Matthew J. Griffiths, Takashi Hibiki and Mamoru Ishii

Abstract Two phase flows in large diameter pipes have immense importance in a wide variety of industrial applications. As a first approximation for the prediction of a two-phase flow and as a beginning for the development of more complex models, the drift-flux model is often used to characterize and predict flows for many geometries and flow conditions. In this chapter, the flow characteristics in flows in large diameter pipes are illustrated based on the experimental data. The flow regimes and their transition criteria are discussed. The existing drift-flux models are summarized, their strengths and weaknesses are noted and the data that can be used to evaluate these models are presented. Based on the flow regime transitions in large diameter pipes, all of the available drift-flux models are evaluated systematically in both low (bubbly) and high (cap and churn-turbulent) void fraction flows. The drift-flux type correlations of Hibiki and Ishii [14] and Kataoka and Ishii [24] are found to be able to give the best predictions for the existing low and high void fraction databases respectively and are recommended for void fraction predictions in flows in large diameter pipes.

Keywords Void fraction · Flow regime · Distribution parameter · Drift velocity · Drift-flux model · Large diameter pipe

---

X. Shen
Research Reactor Institute, Kyoto University, 2-1010 Asashiro-nishi, Kumatori, Sennan, Osaka 590-0494, Japan
e-mail: xzshen@rri.kyoto-u.ac.jp

J. P. Schlegel · S. Chen · S. Rassame · M. J. Griffiths · T. Hibiki (✉) · M. Ishii
School of Nuclear Engineering, Purdue University, 400 Central Drive, West Lafayette, IN 47907-2017, USA
e-mail: hibiki@purdue.edu

J. P. Schlegel
e-mail: schlegelj@mst.edu

L. Cheng (ed.), *Frontiers and Progress in Multiphase Flow I*,
Frontiers and Progress in Multiphase Flow, DOI: 10.1007/978-3-319-04358-6_2,

Nomenclature

A	Area (m^2) or A parameter in Shen et al. [45] drift-flux model (-)
B	A parameter of Shen et al. [45] drift-flux model (-)
B_1	A parameter of Chexal et al. [4] drift-flux model (-)
C_0	Distribution parameter (-)
C_2	A parameter of Chexal et al. [4] drift-flux model (-)
C_3	A parameter of Chexal et al. [4] drift-flux model (-)
C_4	A parameter of Chexal et al. [4] drift-flux model (-)
C_∞	The asymptotic value of C_0 (-)
$C_{D\infty}$	Drag coefficient of a single particle in an infinite medium (-)
D	Diameter (m)
D_b	Bubble diameter (m)
D_H	Hydraulic diameter (m)
F	Placeholder value (-)
g	Gravitational acceleration (m/s^2)
j	Superficial velocity, namely, Volumetric flux (m/s)
K_0	A parameter of Chexal et al. [4] drift-flux model (-)
L	Chexal-Lellouche fluid parameter (-)
$N_{Re\infty}$	Particle Reynolds number in an infinite medium (-)
N_μ	Non-dimensional viscosity number for continuous and disperse phases (-)
$N_{\mu f}$	Non-dimensional viscosity number for liquid and gas phases (-)
r	A parameter of Chexal et al. [4] drift-flux model (-)
r_d	Radius of a particle (m)
Re	Reynolds number (-)
Re_f	Reynolds number for liquid phase only (-)
Re_g	Reynolds number for the gas phase only (-)
$\bar{V}_{gj}$	Mean transport drift velocity (m/s)
V_{gj}^0	A drift velocity parameter of Chexal et al. [4] drift-flux model (m/s)
v	Velocity (m/s)
v_{gj}	Drift velocity (m/s)
z	Height (m)

Greek Letters

α	Void fraction (-)
μ	Dynamic viscosity (Pa s)
ρ	Density (kg/m^3)
$\Delta\rho$	Density difference between the gas and liquid phases (kg/m^3)
ν	Kinematic viscosity (m^2/s)
σ	Surface tension (N-m)

Subscripts/Superscripts

*	Non-dimensional value
+	Non-dimensional value
∞	In an infinite medium

B	Value for bubbly flow
c	Critical value or continuous phase
cap	Cap bubble
f	Liquid phase
g	Gas phase
P	Value for cap-bubbly flow
r	Relative velocity between two phases
S	Value for slug flow
$sphere$	Spherical bubble

Operators

$\langle\rangle$	Area-averaged value
$\langle\langle\rangle\rangle$	Void-fraction-weighted mean value

2.1 Introduction

Two-phase flows are essential in industries ranging from energy generation and petrochemical processing to pharmaceuticals. Two-phase flow systems have even been proposed for atmosphere scrubbing and waste disposal as part of the space program. The reason for this is that two-phase flows are extremely efficient for transferring mass or energy from one location to another and provide a simple method for generating large surface areas to allow chemical reactions between gases and water-soluble compounds.

Two-phase flows are also an essential component in nuclear energy systems. In Pressurized Water Reactors (PWR), the steam generator is a heat exchanger that creates the steam needed to run the turbine. The flow in this region is extremely complex and must be accurately predicted in order to properly predict the characteristics of the reactor during normal operation as well as transient scenarios. Likewise for Boiling Water Reactors (BWR), where steam is generated in the core itself. While this provides a negative reactor feedback, improving safety, it also means that two-phase flows are present in the reactor core. In BWRs which utilize natural circulation to establish liquid flow through the core, two-phase flow also exists in the large-diameter chimney section installed above the core to provide the hydrostatic head necessary for driving the natural circulation. The ability to predict two-phase flows in these rod bundle and large diameter channel regions is essential for predicting the safety performance and power characteristics of these types of reactors.

Two-phase flows are characterized by the existence of one or several interfaces and discontinuities in fluid properties at the interface. The interaction between the two phases causes the interfaces to change continuously, which may result in enhancement or reduction in the efficiency of transfer of energy or mass in the two-phase flows. The interactions between phases are closely associated with the drag forces.

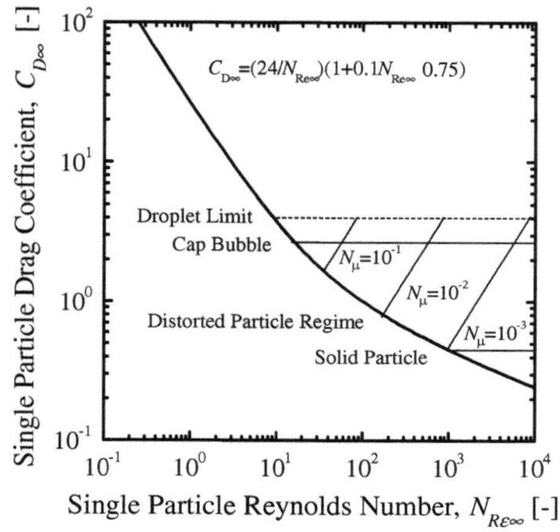

Fig. 2.1 Single-particle drag coefficients [21, 22]

For a single particle moving in an infinite medium the drag coefficient, $C_{D\infty}$, depends on the particle Reynolds number and the viscosity number [21, 22] as shown in Fig. 2.1.

$$N_{\mathrm{Re}\infty} = \frac{2r_d\rho_c|v_{r\infty}|}{\mu_c} \tag{2.1}$$

and

$$N_\mu = \frac{\mu_c}{\left(\rho_c\sigma\sqrt{\frac{\sigma}{g\Delta\rho}}\right)^{1/2}}. \tag{2.2}$$

where r_d, $v_{r\infty}$, μ_c, ρ_c, $\Delta\rho$, σ and g are the particle radius, the relative velocity of a single particle in an infinite medium, continuous phase viscosity, continuous phase density, density difference between the continuous and dispersed phases, surface tension and gravitational acceleration, respectively. Figure 2.1 shows that $C_{D\infty}$ greatly depends on the bubble shape (such as spherical, distorted or cap) and size.

Generally, bubbles are categorized into two groups based on these drag characteristics. Group 1 bubbles are defined as small spherical and ellipsoid (or distorted) bubbles. Group 2 bubbles are larger cap-shaped or irregularly shaped bubbles. These groups are determined based on the drag behavior of each bubble group. Thus this bubble classification provides a clear distinction between two types of bubbles with varying behavior, while bubbles within either group behave in relatively similar fashion to each other.

Materials are transported in industrial systems by tubes and pipes of varying sizes ranging from microchannels to large diameter pipes. The diameters of the tubes and pipes may limit the shape of the bubbles and restrict their ability to move within the channel. Thus the behavior of the bubbles is closely linked with the

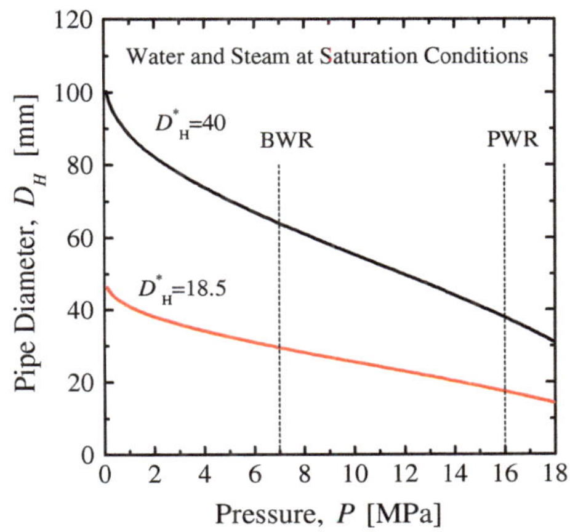

Fig. 2.2 Pipe diameter change with pressure at water and steam saturation states

diameter of the tubes and pipes. Tubes and pipes with non-dimensional hydraulic diameter, defined as

$$D_H^* = \frac{D_H}{\sqrt{\sigma/g\Delta\rho}} \tag{2.3}$$

smaller than about 18.5 are considered 'small diameter' due to the presence of stable Taylor bubbles forming long gas slugs which occupy the entire cross section of the pipe. Pipes with non-dimensional diameter larger than 40 are considered large diameter pipes because Taylor bubbles occupying the entire pipe cross section can no longer be sustained. The intervening range, from non-dimensional diameter of 18.5–40, is a transition region between the two behaviors [37]. Pipe diameters corresponding to $D_H^* = 18.5$ and $D_H^* = 40$ of water and steam at saturation states are shown in Fig. 2.2 for different working pressures, with the operating pressures of Boiling Water Reactors (BWR) and Pressurized Water Reactors (PWR) highlighted.

Prediction of flows in large diameter pipes presents several unique challenges. Surface instability results in the breakup of large Taylor cap bubbles into smaller bubbles, and even these bubbles have highly distorted surfaces. This prevents the formation of stable slug bubbles [24]. The effects of this bubble breakup and surface instability on the flow are varied and complex, affecting all facets of the hydrodynamics of the flow. This makes the prediction of interfacial drag, interfacial structure, generation of turbulence, and other important phenomena difficult at best.

Many modeling efforts in two-phase flows begin with the drift-flux model [58]. Despite being less rigorous than the more detailed two-fluid model, the drift-flux model is extremely important for predicting two-phase flows because of its intrinsic simplicity and because the drift velocity and distribution parameter are directly related to the physical structure of the flow [22]. For this reason, the drift-flux model is a reasonable starting point for understanding and modeling the behavior of two-phase flows in large diameter pipes.

Due to the importance of large diameter pipes in two-phase systems and the challenge that predicting these flows poses, significant effort has already been made to develop the tools necessary to do so. Many researchers have performed hundreds of measurements and developed many drift-flux type models. Many of these models are empirical in nature and, while they are of limited use in predicting the wide range of two-phase flows in large diameter systems, provide valuable insight into the most important factors affecting the flow. Data has been collected in pipes with non-dimensional diameters as high as 220 and in adiabatic air-water and steam-water flows as well as boiling flows.

This chapter will focus on summarizing the most important characteristics of flows in large diameter pipes as well as the drift-flux models available. The characteristics of two-phase flows in large diameter pipes and their effects on the behavior and prediction of the flow will be discussed. Then the available drift flux models will be presented and the major properties, strengths and weaknesses of each will be discussed. Finally the data available for evaluating and analyzing the performance of the drift-flux models will be summarized and analyzed.

2.2 Characteristics of Large Diameter Pipes

In comparison with two-phase flows in small diameter pipes, flows in large diameter pipes are quite different. The differences are summarized as follows.

2.2.1 Surface Instability and Shearing off in Large Pipes

In small diameter channels, the channel walls restrict the growth of large bubbles, resulting in the formation of gas slugs. Because the diameter of these gas slugs is relatively small, the upper bubble surface remains stable and distortions in the bubble surface, which may be caused by turbulent fluctuations or by Kelvin–Helmholtz instability, are quickly damped. In large diameter pipes however, the diameter of the gas slugs can become very large. In this case, the shape of the upper bubble surface as not so restrained. Because of this when distortions to the bubble surface occur Taylor instabilities result in growth of the disturbance. This causes the bubble to break up, as shown in Fig. 2.3. Generally when discussing bubble breakup mechanisms, this mechanism is described as 'surface instability'. The limited ability of bubbles to grow caused by this surface instability, resulting in the absence of large gas slugs, is the defining feature of large diameter channels.

The increased prevalence of bubble breakup due to surface instability results in some significant differences between flows in large and small diameter channels. First, in place of large gas slugs flows in large diameter channels are composed of either Taylor cap bubbles (at moderate void fractions) or unstable churn-turbulent bubbles (at high void fractions). This change affects the flow in a variety of ways.

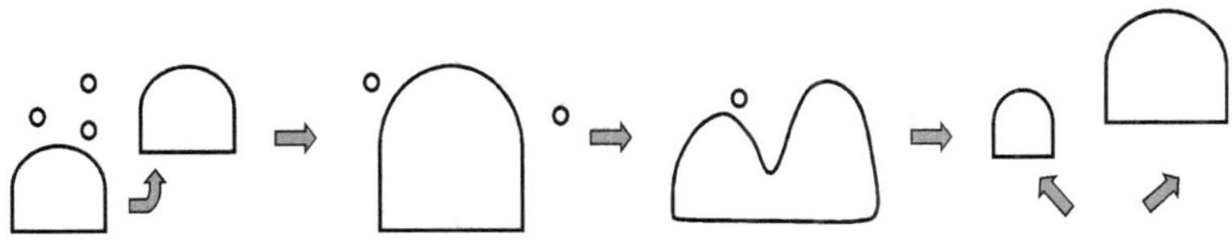

Fig. 2.3 Formation and breakup of cap bubbles larger than the maximum stable bubble size

Fig. 2.4 Enhanced turbulence and shearing-off of small bubbles at base of cap bubble

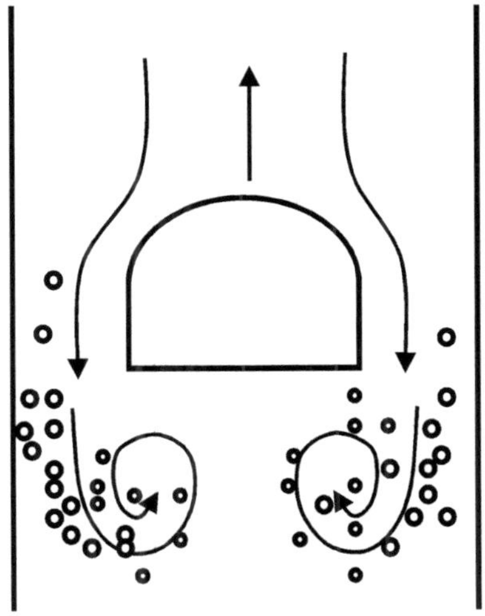

The increased number of Group 2 bubbles results in additional interfacial surface area in the flow. This can result in dramatically increased bubble-induced turbulence. Additionally, the presence of an increased number of cap bubbles results in many additional locations where small bubbles can be broken off from the base of larger bubbles, as shown in Fig. 2.4. This breakup mechanism, which is termed 'shearing-off,' may become the most significant source of small, dispersed bubbles in large diameter channels and result in a higher proportion of the void being composed of smaller Group 1 bubbles [9].

In spite of these differences, many of the other ways in which bubbles interact are very similar. In general, bubbles may coalesce by randomly bumping into each other within the flow. This type of coalescence, termed 'random collision,' occurs mainly with small bubbles and depends on the bubble concentration and size of small bubbles, as well as the strength of turbulence in the flow. Both the bubble concentration and size of Group 1 bubbles are not strongly dependent on the pipe size. Generally the turbulence characteristics of flows large diameter channels are quite different from those in small diameter channels, but this is generally accounted for using constitutive models. The second major type of bubble coalescence occurs when bubbles are caught up in the wake of a preceding bubble and run into each other. This is called 'wake entrainment' and is determined by the relative velocity of the bubbles, their geometry, and the concentration of the bubbles. Again, this process is very similar in both large and small diameter channels, and the differences that do exist can be accounted for in the models for predicting relative velocity and concentration. The most significant cause of bubble breakup is the interaction of bubbles with turbulent eddies in the liquid. This mechanism is called ‘turbulent impact.’ Generally this mechanism is not very strong for Group 2 bubbles, but for Group 1 bubbles can be very important. This mechanism is dependent on the Group 1 bubble properties, which do not change much as the channel size increases, as well as the turbulence properties. Turbulence characteristics can be very different in large and small diameter channels, however this does not change the behavior of the mechanism and can be accounted for in the models for predicting turbulence. Thus for all of these types of bubble interactions, the actual physics is very similar in the various pipe sizes and the differences can be accounted for by using the correct models for predicting relative velocity, turbulence, and other important parameters [22].

2.2.2 Flow Regimes

In small diameter pipes, the flow regimes typically noted include bubbly, slug, churn and annular flows [28]. As discussed in the previous section however, stable slug bubbles cannot exist in large diameter channels. This means that flow regime maps developed by various researchers for small diameter pipes cannot be applied to large diameter channels. This has been confirmed by observations from many sources [31, 38, 44, 59].

The first investigation which defined the flow regimes present in large diameter pipes in detail was that of [31], where flow regime characteristics were investigated in a pipe with diameter of 0.2 m. The flow regimes were classified as follows:

- Undisturbed bubbly: composed of very small numbers of small, spherical bubbles which do not show significant lateral movement.
- Agitated bubbly: composed of small, spherical bubbles which show some mixing but limited coalescence or breakup.
- Churn bubbly: composed of many small, spherical bubbles and characterized by very significant mixing and prevalent bubble breakup and coalescence.
- Churn slug: composed of many small, spherical bubbles as well as large but relatively stable cap-shaped bubbles.
- Churn-froth: composed of many small, spherical bubbles as well as unstable large bubbles that undergo frequent coalescence and breakup.

The results of their flow regime identification efforts can be seen in Fig. 2.5. These flow regimes classifications were determined by the visual observation of the researchers. Unfortunately, the flow conditions in this study were restricted to gas velocities no higher than 4.7 m/s. This means that the transition to annular flow was not able to be evaluated. Annular flow is characterized by a solid gas core surrounded by a liquid film. This gas core may also contain liquid droplets.

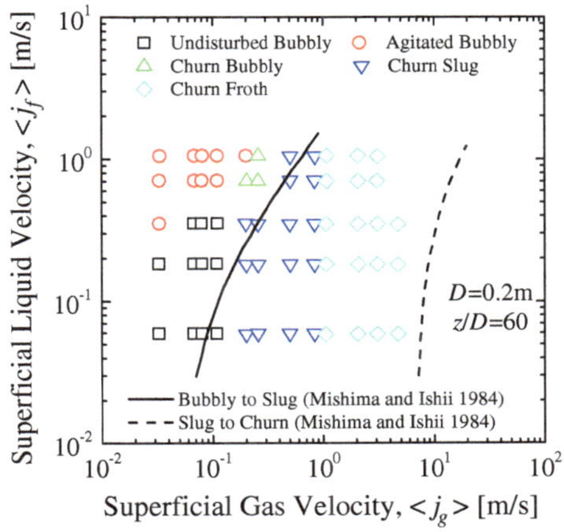

Fig. 2.5 Flow regimes observed by Ohnuki and Akimoto [31]

In view of the difficulties in visualizing the internal flow structure in a large diameter pipe, Smith [49] and Schlegel et al. [38] observed the flow regimes by means of electrical impedance void meters in pipes of various diameters. In these studies, the flow regimes were re-evaluated based on the transport properties of the bubbles. Bubbly flow includes the undisturbed bubbly, agitated bubbly and churn bubbly regimes defined by Ohnuki and Akimoto [31]. Cap-bubbly or cap-turbulent flow is identical to the churn-slug flow regime, while churn-turbulent flow is identical to the churn-froth flow regime. It is recommended that these flow regimes be used when discussing two-phase flows and in developing models, as these flow regimes were determined based on the transport properties of bubbles, namely bubble drag and relative velocity, and these transport properties determine the behavior of the bubbles and therefore the properties of two-phase flows. Shen et al. [43, 44] reported bubbly flow, churn flow and slug flow in a 0.2 m diameter pipe. In their case, slug flow is defined as intermittent upward-flowing large cap bubbles and is identical to cap-turbulent flow.

In small diameter pipes, the flow regime transition occur rather abruptly as the mechanisms which cause the change in flow regime occur under very specific conditions. For large diameter pipes however, all of these researchers have noted that the transitions between flow regimes are very gradual [45]. This is because in large diameter channels, Group 2 bubbles behave in a much more random fashion than for flows in small diameter channels, and the mechanisms which cause the transitions between flow regimes reflect this behavior.

2.2.3 *Effect of Inlet Conditions on Flow*

Another major characteristic of two-phase flows in large diameter pipes is the effect of the inlet condition on the flow pattern [13]. For small diameter channels the gas injection method has little effect on the flow, but in large diameter pipes the

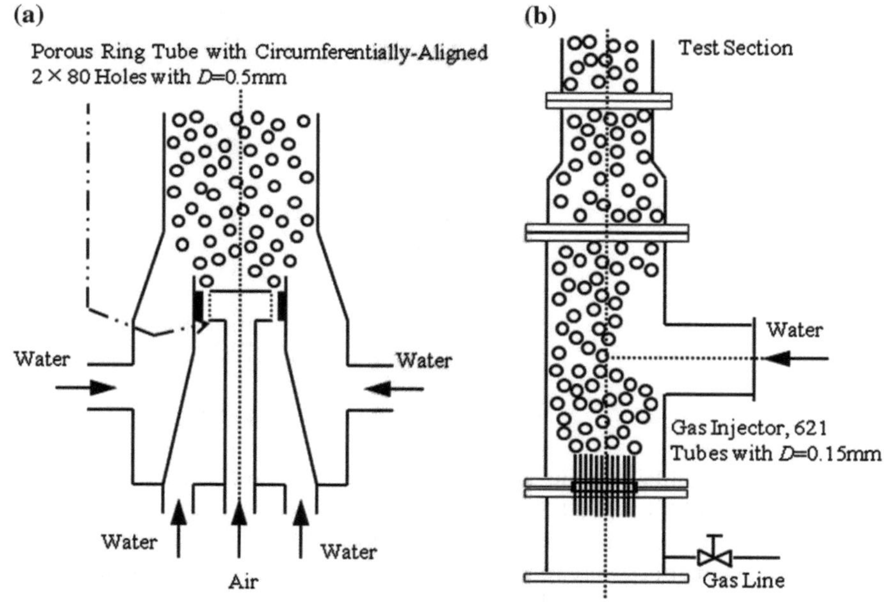

Fig. 2.6 Uniformly distributed bubble injection mixer. **a** Shear-cutting liquid and gas mixer. **b** Needle-injecting liquid and gas mixer

effect is pronounced for relatively low flow rates. At low void fraction conditions, almost uniformly distributed small, spherical bubbles are obtained by using the typical liquid and gas inlet mixers such as the shear-cutting liquid and gas mixer of Shen et al. [44] and the needle-injecting liquid and gas mixer of Hibiki and Ishii [13] shown in Fig. 2.6a, b, respectively. At identical gas and liquid injection rates, cap bubbles at the inlet are observed for bubble injection using a horizontal section in Fig. 2.7 [13] even though the condition is typically considered bubbly flow. Bubbles injected as large cap bubbles do not break up easily and are typically much less affected by turbulence because they are significantly larger than the turbulent eddy sizes which contain most of the turbulent kinetic energy. This means that cap bubbles can be found even at very low void fractions (<0.1) when the gas is introduced by using the bubble injection mixer with a horizontal section. This can significantly affect the flow under these conditions, with very different flow behavior at low void fractions under different injection conditions.

2.2.4 Relative Velocity in Large Diameter Pipes

Large-volume cap bubbles can be found in both small and large diameter channels. In the small diameter case, these appear as slug bubbles which occupy the entire pipe diameter. Generally, the slug bubble length is less significant when determining how

Fig. 2.7 Bubble injection mixer with elbow horizontal section

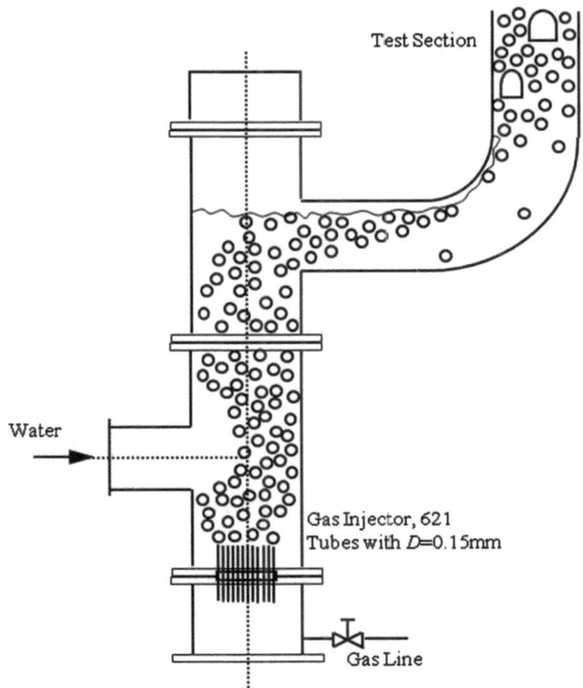

the bubble moves in a two-phase mixture and the dynamics of the flow around the nose and tail of the bubble almost entirely govern their motion. Thus the small diameter of the bubble nose and the effect of wall drag on the bubble limit the rise velocity of slug bubbles in the liquid. This effect can be seen for smaller pipe diameters in Fig. 2.8, which shows the drift velocity collected by Kataoka and Ishii [24] for several pipe diameters. As the pipe size increases the shape of the bubble nose changes and the effect of wall drag is reduced, leading to higher bubble relative velocity.

In large diameter pipes however, the figure shows that the drift velocity reaches a maximum and then remains constant. In larger channels the channel size is larger than the maximum stable bubble. When this occurs, the mechanics governing the relative velocity are no longer dependent on the channel size as the channel no longer restricts bubble motion. Additionally, the additional drag on large bubbles caused by the wall is no longer a factor because the Group 2 bubbles are generally no longer in contact with the wall. Because of this, the relative velocity is determined entirely by the bubble hydrodynamics without the effect of the wall and becomes independent of the pipe diameter [12, 43].

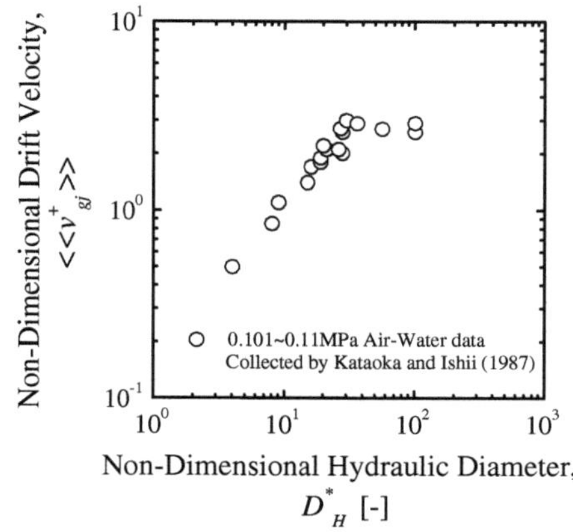

Fig. 2.8 Drift velocity change with pipe diameter

2.2.5 Profiles of Local Flow Parameters

Turbulence plays a vital role in determining the local void fraction and velocity profiles. In general, turbulence is generated in regions where a velocity gradient exists. For single-phase flows, this is in the region near the wall of the flow channel. In small diameter channels, this is very similar for two-phase flows as shown by Liu and Bankoff [26]. For large diameter channels, single-phase flows show very similar turbulence characteristics to flows in small diameter channels. On the other hand, while the existing data regarding turbulence characteristics in large diameter channels is admittedly limited, it indicates that two-phase flows are very different.

Based on the work of Serizawa and Kataoka [41], two-phase flows introduce bubble-induced turbulence as well as turbulence absorption at the interface. Absorption of turbulence at the interface is caused by distortions in the bubble surface. These distortions absorb energy because of surface tension, which results in a small reduction in turbulence. Bubble-induced turbulence is dependent on the relative velocity between the phases, drag force, and interfacial area concentration. For bubbly flows where only Group 1 bubbles are present this bubble-induced turbulence is similar in magnitude to the absorption of turbulence at the interface and can, under some conditions, actually result in reduction turbulence as compared to single-phase flows. Group 2 bubbles however have very different characteristics than Group 1 bubbles. While the relative velocity is similar, the drag force is much larger for Group 2 bubbles and the interfacial area concentration is much smaller. The net result of these two effects is that turbulent kinetic energy production from Group 2 bubbles is about an order of magnitude higher than that from Group 1 bubbles. Thus the bubble-induced turbulence created by Group 2 bubbles dominates in cap-bubbly and churn-turbulent flows for all types of flow channels.

This leads to the differences in turbulence characteristics for flows in small and large diameter channels. Generally, the Group 1 bubbles are similar in size and behavior regardless of channel size. However, based on the work of Ohnuki and Akimoto [32], the length scale at which turbulence is produced is very important. Turbulent eddies significantly larger than the bubble size tend to carry more energy and carry bubbles along, while those significantly smaller do not appreciably affect the bubbles. In large diameter channels, turbulence is produced at much larger scales than in smaller channels. This means that the turbulent kinetic energy and turbulent fluctuations may be higher in large channels even though the velocity gradient near the pipe wall may be smaller, resulting in significant changes to the bubble distribution and small changes in bubble size and behavior. As discussed in Sect. 2.2.1, large diameter pipes cannot sustain stable slug bubbles. In the cap-bubbly and churn-turbulent flow regimes, this leads to the existence of many smaller cap bubbles within the flow rather than one large gas slug. This drastically increases the interfacial area concentration of Group 2 bubbles, leading to a corresponding increase in the production of turbulent kinetic energy from these bubbles. Because of these two effects it is expected that flows in large diameter channels will have significantly stronger turbulent fluctuations than flows in small diameter pipes across nearly all flow conditions, but especially at relatively low liquid flow rates. This is illustrated in the data of Ohnuki and Akimoto [31].

These changes in the turbulence characteristics for large diameter channels result in very different flow behavior. In small diameter pipes, the lift force causes small spherical bubbles within a certain range of bubble diameter to concentrate near the pipe wall at low void fractions. This phenomenon is known as wall-peaking. The increased turbulence as well as larger scale of turbulence in large diameter channels results enhanced turbulent mixing of small, spherical bubbles. Because of this enhanced mixing as well as a reduction in lift force due to a smaller near-wall velocity gradient, the bubbles are distributed more evenly throughout the flow channel, resulting in elimination or reduction of the wall-peaking effect. This effect is shown in Fig. 2.9 using void profile data from experiments in 50.8 mm [15] and 200 mm [47] diameter channels. The figure shows the void fraction profile for very similar flow conditions and axial locations, with the only significant difference being the size of the channel. Even so, the void fraction profiles are quite different. Similar trends are shown in experiments performed by Smith et al. [51] in 101 and 152 mm diameter channels and by [39] in channels with diameter of 152 and 203 mm.

Large cap bubbles also have many other effects on the velocity profile in large channels. At lower flow rates, large bubbles tend to collect in higher-velocity regions within the pipe—namely the center of the pipe. Because of this, the velocity in this region may be enhanced at higher void fractions. This effect can be noted in the data of Ohnuki and Akimoto [30, 31] as well as the data of Smith et al. [51] Sun et al. [54], Shen et al. [59] and Schlegel et al. [39] in the gas velocity profiles at various flow conditions which show a center peak in the gas velocity profile which becomes more prominent as the void fraction increases.

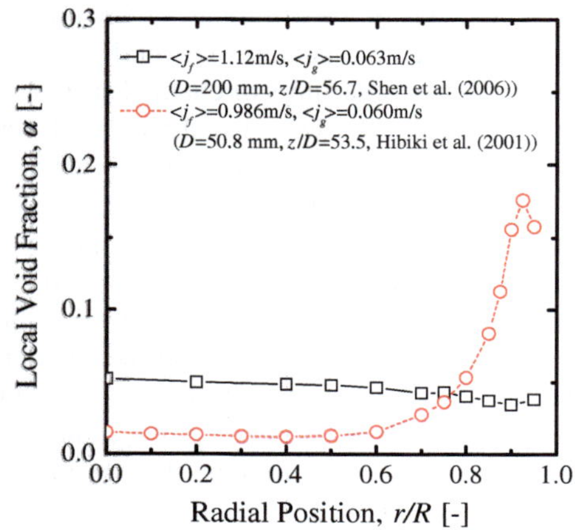

Fig. 2.9 Wall-peaking effect comparison between small and large diameter pipes

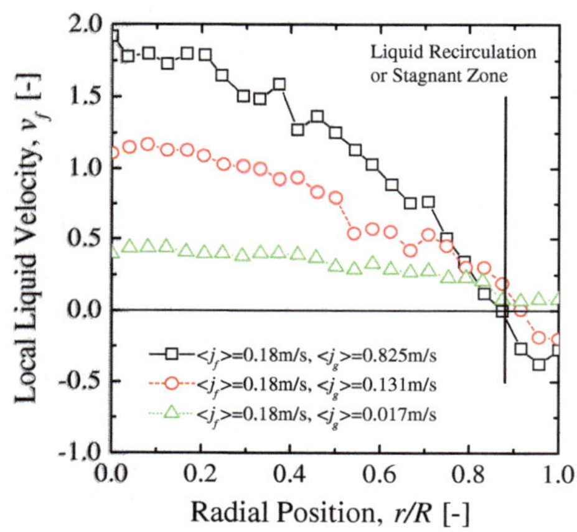

Fig. 2.10 Radial local liquid velocity in a large diameter pipe from Ohnuki et al. [33]

When the liquid velocity is very small, this effect can have some interesting consequences. As noted by Hills [16], when the liquid velocity is very small compared to the bubble rise velocity, the liquid is pulled along by the large bubbles in the center of the channel more quickly than the liquid is flowing near the channel wall. In this case, to maintain the total liquid volumetric flow rate, the liquid tends to move out of the pipe center toward the channel wall where it can be stagnant or even move against the total flow direction. This is illustrated by the data of Ohnuki et al. [33] in Fig. 2.10. The figure shows that at low gas velocities when cap bubbles do not exist, the liquid velocity is relatively flat. At higher gas velocities where larger bubbles begin to dominate, the velocity profile develops a strong center peak with negative local velocities near the channel wall. Based on the data from this and other studies, this effect tends to decrease as the liquid flow rate increases and for average liquid velocities higher than about 0.5 m/s this recirculation region near the channel wall is largely absent.

2.3 Flow Regime Transition Criteria

As many existing correlations are applicable only for certain flow patterns, accurate flow pattern maps which show the conditions for which certain flow patterns are present are key. For flows in small pipes, theoretically-based flow regime maps have been developed by Taitel et al. [55] and Mishima and Ishii [28], but the applicability of these models to large pipes has not been thoroughly investigated. Schlegel et al. [38] have developed a flow regime map for large diameter channels which has been partially validated using a large database of flow regime identification data.

Before defining flow regime transitions, it is first necessary to define the flow regimes. Based on experiment and as described in Sect. 2.2.2, there are four flow regimes with unique transport characteristics present in large diameter channels. The first is bubbly flow, which consists largely of small, spherical bubbles and occurs at lower void fraction. The second flow regime is cap-turbulent flow, which is characterized by large Taylor bubbles moving through the liquid phase. This results in significant agitation of the flow and causes the liquid and smaller bubbles to move in very chaotic fashion, although the cap bubbles are still relatively stable. The third flow regime is churn-turbulent flow, which is characterized by highly unstable large bubbles which coalesce and break up with extremely high frequency due to the proximity of large bubbles to each other and surface instability. This results in a very chaotic flow pattern with irregular bubble shapes. These three flow regimes approximately correspond to the agitated bubbly, churn slug, and churn froth flow regimes described by Ohnuki and Akimoto [31]. The final flow regime is annular/mist flow, which is characterized by a gas core containing liquid droplets and surrounded by a liquid film [38].

Both theoretical flow regime maps for small diameter channels [55, 28] considered packing of bubbles and a sudden increase in the coalescence rate as the mechanism for the transition from bubbly flow to slug flow. Taitel and Bornea [55] postulated a cubic bubble lattice while Mishima and Ishii [28] postulated a tetrahedral bubble lattice, shown in Fig. 2.11, resulting in predictions that the transition would occur at void fractions of 0.25 and 0.30, respectively. This mechanism is very likely the mechanism which drives the transition from bubbly to cap-turbulent flow in large diameter channels, and so it is expected that the transition boundary will be similar. In fact, experimental studies have shown that the transition region tends to lie between these two values of void fraction based on changes in the drift-flux parameters and on flow regime identification studies. Thus it is recommended that $\langle\alpha\rangle = 0.3$ be used as the flow regime transition criteria from bubbly to cap-turbulent flow in large diameter channels, but one must keep in mind that the transition begins at lower void fractions, about 0.25, and be cautious when applying models in this transition region [38].

In large diameter channels, Taylor bubbles occupying the entire diameter of the channel simply cannot exist for an extended length of time due to instability in the upper surface of such large bubbles. Therefore, stable slug bubbles cannot exist

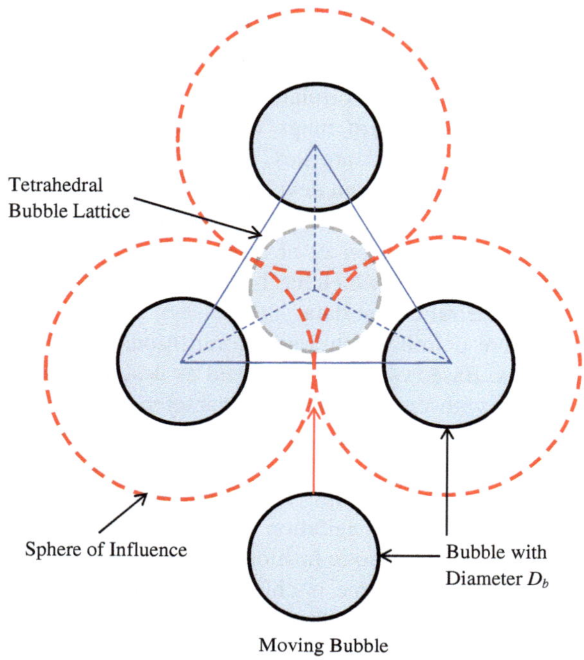

Fig. 2.11 Maximum packing of spherical bubbles for transition to cap/slug flow

and the transition criteria from slug flow to churn-turbulent flow in small-diameter studies cannot be applied for the transition from cap-turbulent to churn-turbulent flow in large diameter channels. Instead, Schlegel et al. [38] proposed that the mechanism for the transition is the onset of rapid formation and disintegration of Taylor bubbles larger than the bubble size limit. This occurs when Taylor bubbles achieve the maximum packing criteria in similar fashion to bubbly flows. This process is illustrated for Taylor bubbles in Fig. 2.12. The ratio of Taylor bubble volume to the volume of the tetrahedron is approximately 0.3. Thus this should be the cap-bubble fraction at the transition to churn-turbulent flow. If the spherical bubble fraction of the remaining liquid volume is assumed to be 0.30 as in the bubbly flow analysis, then the total void fraction at the transition to churn-turbulent flow is [38]

$$\langle \alpha \rangle = \langle \alpha_{Cap} \rangle + \langle \alpha_{Sphere} \rangle \left(1 - \langle \alpha_{Cap} \rangle\right) = 0.51 \tag{2.4}$$

The transition to annular flow which is applicable to large diameter pipes is the entrainment condition postulated by Mishima and Ishii [28]. Interestingly, this is the same mechanism hypothesized by Taitel and Bornea [55], and the resulting transition boundaries for the two studies have almost identical formulations. In fact, the only difference is that in Mishima and Ishii's formulation, the constant 3.1 in Taitel and Bornea's model is replaced by the viscosity number, which scales the effect of the liquid viscosity on the flow and is equal to approximately 3.38 for air-

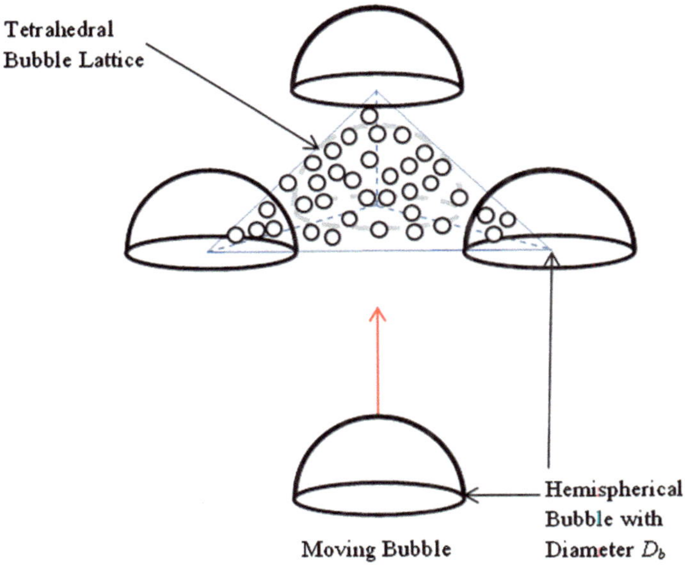

Fig. 2.12 Maximum packing of cap bubbles for transition to churn-turbulent flow

Table 2.1 Flow regime transition criteria for large diameter pipes

Flow regime transition	Criteria
From bubble to cap-turbulent flow	$\langle\alpha\rangle = 0.30$
From cap-turbulent to churn-turbulent flow	$\langle\alpha\rangle = 0.51$
From churn-turbulent flow to annular/mist flow	$\langle j_g\rangle = \left(\frac{\sigma g \Delta\rho}{\rho_g^2}\right)^{1/4} N_{\mu f}^{-0.2}$

water flows at atmospheric conditions. For this reason, the transition boundary for droplet entrainment proposed by Mishima and Ishii [28] is suggested for use in large diameter pipes to ensure generality [38].

The flow regime transition criteria proposed by Schlegel et al. [38] are summarized in Table 2.1. Based on these proposed transition criteria, the flow regime map is shown in Fig. 2.13 including the flow regime identification results of Schlegel et al. [38] using impedance void meter measurements and a self-organized neural network to determine the flow regime of each condition. As the figure shows, the proposed transition from bubbly to cap-turbulent flow agrees almost exactly with the data as does the proposed transition from cap-turbulent to churn-turbulent flow. Additionally, Figs. 2.14 and 2.15 show the comparison of these flow regime transition criteria with the observations of Ohnuki and Akimoto [32] and Smith [49]. Ohnuki and Akimoto [32] used visual observation and categorization to determine the flow regime. In Fig. 2.14, the undisturbed bubbly and agitated bubbly flow conditions have been combined as 'Bubbly' and the observations at the test

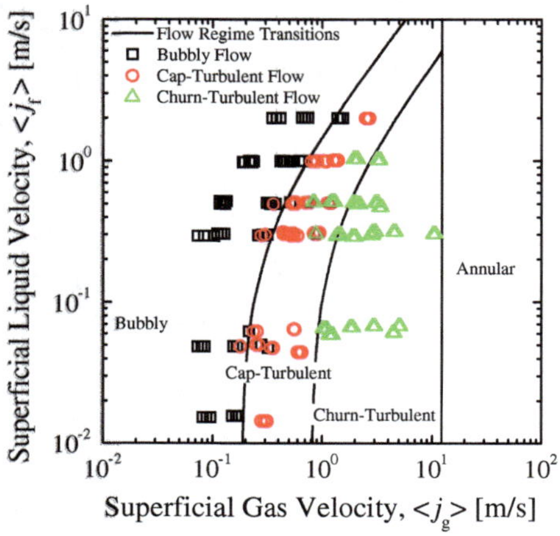

Fig. 2.13 Flow regime mapping and identification of Schlegel et al. [38]

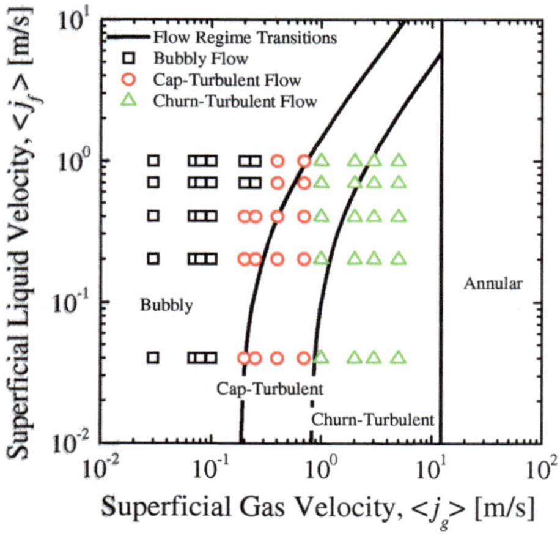

Fig. 2.14 Flow regime identification results of Ohnuki and Akimoto [31]

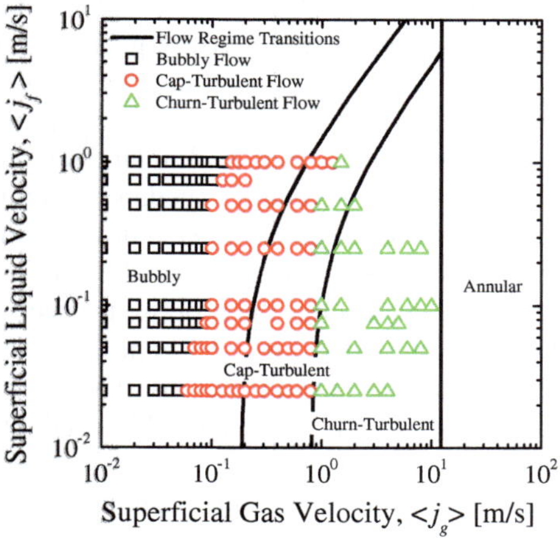

Fig. 2.15 Flow regime identification results of Smith [49]

section exit are used. Smith [49] used a neural network to categorize the flow regime based on data collected by impedance void meters to generate the results shown in Fig. 2.15. It should be noted that this result shows that cap-turbulent flow begins at significantly lower void fraction than the other studies. This may be caused by the injection method, which likely resulted in cap-bubbly injection when the gas flow rate reached values higher than 0.1 m/s. Otherwise, both experimental studies confirm the applicability of the flow regime transitions developed by Schlegel et al. [38] for large diameter channels.

2.4 Drift-Flux Models

2.4.1 Formulation of the Drift-Flux Model

The drift-flux model is one of the most practical and accurate models for two-phase-flow analysis. The model takes into account the effects of non-uniform velocity and void fraction profiles as well as the effect of the local relative velocity between phases. It has been utilized to solve many engineering problems involving two-phase-flow dynamics. In particular, its application to forced convection systems has been quite successful. The one-dimensional drift-flux model was derived by averaging the local drift velocity over the channel cross-section [58]. The model is given as

$$\frac{\langle j_g \rangle}{\langle \alpha \rangle} = \langle\langle v_g \rangle\rangle = C_0 \langle j \rangle + \langle\langle v_{gj} \rangle\rangle \tag{2.5}$$

where $\langle\rangle$ denotes the area average of a quantity, F, over the cross-sectional area (A) of the flow path, which is defined mathematically as

$$\langle F \rangle = \frac{1}{A} \int_A F dA \tag{2.6}$$

and α and j are the void fraction and the mixture volumetric flux, respectively. $\langle\langle v_g \rangle\rangle$ is the void-fraction-weighted mean gas velocity. The distribution parameter, C_0, and the void-fraction-weighted mean drift velocity, $\langle\langle v_{gj} \rangle\rangle$, are respectively defined by the following equations,

$$C_0 = \frac{\langle \alpha j \rangle}{\langle \alpha \rangle \langle j \rangle} \tag{2.7}$$

$$\langle\langle v_{gj} \rangle\rangle = \frac{\langle v_{gj} \alpha \rangle}{\langle \alpha \rangle} \tag{2.8}$$

where v_{gj} is the local drift velocity of gas phase defined as

$$v_{gj} = v_g - j = (1 - \alpha)(v_g - v_f) \tag{2.9}$$

The v_g and v_f in the above equation are the gas and liquid velocities, respectively.

The void-fraction-weighted mean gas velocity, $\langle\langle v_g \rangle\rangle$, and the cross-sectional mean mixture volumetric flux, $\langle j \rangle$, are easily obtainable parameters in experiments. Therefore, Eq. (2.5) suggests a plot of $\langle\langle v_g \rangle\rangle$ versus $\langle j \rangle$. An important characteristic of such a plot is that, for two-phase flow regimes with fully developed void and velocity profiles, the data points cluster around a straight line. The value of the distribution parameter, C_0, can be obtained indirectly from the slope of the line, whereas the intercept of this line with the void-fraction-weighted mean gas velocity axis can be interpreted as the void-fraction-weighted mean drift velocity, $\langle\langle v_{gj} \rangle\rangle$. As the recent development of local sensor techniques [25–27, 40, 46, 61, 62] enables the measurement of the local flow parameters in bubbly flow, including void fraction as well as gas and liquid velocities, the values of C_0 and $\langle\langle v_{gj} \rangle\rangle$ in bubbly flow can be determined directly by Eqs. (2.7) and (2.8) from experimental measurement of the local flow parameters.

It should be mentioned here that sometimes the following non-dimensional parameters are introduced to nondimensionalize the drift-flux model:

$$\langle j_g^+ \rangle = \frac{\langle j_g \rangle}{\left(\frac{\sigma g \Delta\rho}{\rho_f^2}\right)^{1/4}} \tag{2.10}$$

$$\langle j_f^+ \rangle = \frac{\langle j_f \rangle}{\left(\frac{\sigma g \Delta\rho}{\rho_f^2}\right)^{1/4}} \tag{2.11}$$

$$\langle j^+ \rangle = \frac{\langle j \rangle}{\left(\frac{\sigma g \Delta\rho}{\rho_f^2}\right)^{1/4}} \tag{2.12}$$

$$\langle\langle v_g^+ \rangle\rangle = \frac{\langle\langle v_g \rangle\rangle}{\left(\frac{\sigma g \Delta\rho}{\rho_f^2}\right)^{1/4}} \tag{2.13}$$

$$\langle\langle v_{gj}^+ \rangle\rangle = \frac{\langle\langle v_{gj} \rangle\rangle}{\left(\frac{\sigma g \Delta\rho}{\rho_f^2}\right)^{1/4}} \tag{2.14}$$

The non-dimensionalized drift-flux model is expressed as

$$\frac{\langle j_g^+ \rangle}{\langle \alpha \rangle} = \langle\langle v_g^+ \rangle\rangle = C_0 \langle j^+ \rangle + \langle\langle v_{gj}^+ \rangle\rangle. \tag{2.15}$$

2.4.2 Summary of Drift-Flux Correlations for Large Diameter Pipes

Several important studies regarding the development of drift-flux correlations for large diameter pipes are summarized in this section. The details for each model can be found in Table 2.2.

Hills [16] performed tests with a large pipe with an inner diameter of 0.150 m and height of 10.5 m for gas superficial velocities of 0.070–3.5 m/s and liquid superficial velocities of 0–2.7 m/s. Hills developed a drift flux type correlations based on his own experimental data. However, the effect of physical properties on the distribution parameter and the drift velocity are not included in this correlation. Therefore, Hills's correlation may not be applicable to other fluid systems such as high pressure steam-water flow conditions.

Ishii's model [20] was developed as a comprehensive model for small-diameter channels and represents the starting point for many later models. An expression for the distribution parameter was developed semi-empirically, and models for the drift velocity in each flow regime were developed that account for dependence on void fraction, fluid properties and pipe diameter. Unfortunately, as the presence of slug bubbles was assumed and the effect of inlet conditions were not accounted for in the models for bubbly and slug flow, this model may not be completely applicable to large diameter pipe flows.

Shipley [48] conducted experiments with a large pipe of an inner diameter of 0.457 m and height of 5.64 m. A drift flux correlation was developed based on this data. It is noted that the second term in the right hand side of his correlation, corresponding to the drift velocity, can become infinitely large for a very large diameter pipe. This seems not to be physically realizable.

Clark and Flemmer [5] performed tests with a large pipe of an inner diameter of 0.10 m. Mixture volumetric fluxes are ranged from 0.7 to 2.7 m/s. A drift flux correlation was developed based on their own test data. However, their correlation did not take into account the effect of physical properties on the distribution parameter. Consequently, Clark and Flemmer's correlation may not be applicable to other fluid systems such as high pressure steam-water flow conditions. Clark and Flemmer also developed another modified drift-flux type correlation in 1986. It should be noted here that Clark and Flemmer [6] did not consider the effect of physical properties on the distribution parameter and the drift velocity in their

Table 2.2 Drift-flux models for large diameter channels

Researchers	Disribution parameter, C_0 (-)	Drift velocity, $\langle\langle v_{gj}\rangle\rangle$ (m/s)	Applicable range
Ishii [20]	$1.2-0.2\sqrt{\rho_g/\rho_f}$	$\sqrt{2}\left(\frac{\sigma g\Delta\rho}{\rho_f^2}\right)^{1/4}(1-\langle\alpha\rangle)^{1.75}$	Bubbly
	$1.2-0.2\sqrt{\rho_g/\rho_f}$	$0.35\sqrt{\frac{gD_H\Delta\rho}{\rho_f}}$	Slug
	$1.2-0.2\sqrt{\rho_g/\rho_f}$	$\sqrt{2}\left(\frac{\sigma g\Delta\rho}{\rho_f^2}\right)^{1/4}$	Churn
	$\bar{V}_{gj}\equiv(C_0-1)\langle j\rangle+\langle\langle v_{gi}\rangle\rangle=\frac{1-\langle\alpha\rangle}{\langle\alpha\rangle+4\sqrt{\rho_g/\rho_f}}\left[\langle j\rangle+\sqrt{\frac{\Delta\rho gD(1-\langle\alpha\rangle)}{0.015\,\rho_f}}\right]$		Annular
Hills [16]	$1.35/\langle j\rangle^{0.07}$	0.24	$\langle j_f\rangle>0.3$
	1.0	$(1-\langle\alpha\rangle)\left(0.24+4.0\langle\alpha\rangle^{1.72}\right)$	$\langle j_f\rangle<0.3$
Shipley [48]	1.2	$\left\{0.24+0.35\left(\frac{\langle j_g\rangle}{j}\right)^2\sqrt{gD\langle\alpha\rangle}\right\}$	
Clark and Flemmer [5]	$0.934(1+1.42\langle\alpha\rangle)$	$1.53\left(\frac{\sigma g}{\rho_f}\right)^{1/4}$	
Clark and Flemmer [6]	$(1.95\langle j_g\rangle+0.93\langle j_f\rangle)/\langle j\rangle$	0.25	
Murase et al. [29]	$\left\{1.2-0.2\sqrt{\rho_g/\rho_f}\right\}\{1-\exp(-18\langle\alpha\rangle)\}$	$\langle\alpha_c\rangle=0.588-1.817\sqrt{\frac{\rho_g}{\rho_f}}+2\left(\sqrt{\frac{\rho_g}{\rho_f}}\right)-3.34\left(\sqrt{\frac{\rho_g}{\rho_f}}\right)^{3/2}$	$\langle\alpha\rangle<\langle\alpha_c\rangle$
	$1.0+0.2\left(1-\sqrt{\rho_g/\rho_f}\right)\left\{1-\left(\frac{\langle\alpha\rangle-\langle\alpha_c\rangle}{1-\langle\alpha_c\rangle}\right)^2\right\}$		$\langle\alpha\rangle>\langle\alpha_c\rangle$
		$7.1\left(D_H^*\right)^{1/8}\left(\frac{\sigma g\Delta\rho}{\rho_f^2}\right)^{1/4}\left(\frac{1}{\langle\alpha\rangle}-C_0\right)\langle\alpha\rangle^{1.6}$	$\langle\alpha\rangle<0.33$
		$16.4\left(D_H^*\right)^{1/8}\left(\frac{\sigma g\Delta\rho}{\rho_f^2}\right)^{1/4}\left(\frac{1}{\langle\alpha\rangle}-C_0\right)\langle\alpha\rangle\left\{-\ln\left(1-\langle\alpha\rangle^{1.45}\right)\right\}$	$\langle\alpha\rangle>0.33$

(continued)

Table 2.2 (continued)

Researchers	Disribution parameter, C$_0$ (-)	Drift velocity, $\langle\langle v_{gj}\rangle\rangle$ (m/s)	Applicable range
Ishii and Kocamustafaogullari [23]	Not given	$0.54\sqrt{\frac{gD_H\Delta\rho}{\rho_f}}$	$D_H^* \le 30$
	Not given	$3.0\left(\frac{\sigma g\Delta\rho}{\rho_f^2}\right)^{1/4}$	$D_H^* \ge 30$
Hirao et al. [17]	$1.2 - 0.2\sqrt{\rho_g/\rho_f}$	$0.52\sqrt{\frac{gD\Delta\rho}{\rho_f}}$	
Kataoka and Ishii [24]	$1.2 - 0.2\sqrt{\rho_g/\rho_f}$	$V_{gj}^+ = 0.0019(D_H^*)^{0.809}\left(\frac{\rho_g}{\rho_f}\right)^{-0.157} N_{\mu f}^{-0.562}$	$N_{\mu f} \le 2.25\times10^{-3}$ $D_H^* \le 30$
	$1.2 - 0.2\sqrt{\rho_g/\rho_f}$	$V_{gj}^+ = 0.030\left(\frac{\rho_g}{\rho_f}\right)^{-0.157} N_{\mu f}^{-0.562}$	$N_{\mu f} \le 2.25\times10^{-3}$ $D_H^* \ge 30$
	$1.2 - 0.2\sqrt{\rho_g/\rho_f}$	$V_{gj}^+ = 0.92\left(\frac{\rho_g}{\rho_f}\right)^{-0.157}$	$N_{\mu f} > 2.25\times10^{-3}$
Hibiki and Ishii [14]	$\exp\left\{0.475\left(\frac{j_g^+}{j^+}\right)^{1.69}\right\}\left(1-\sqrt{\frac{\rho_g}{\rho_f}}\right)+\sqrt{\frac{\rho_g}{\rho_f}}$	$V_{gj}^+ = V_{gj,B}^+\exp\left(-1.39\langle j_g^+\rangle\right)$ $+V_{gj,P}^+\left\{1-\exp\left(-1.39\langle j_g^+\rangle\right)\right\}$	Bubbly flow $\langle\alpha\rangle < 0.3$ $0 \le \frac{\langle j_g^+\rangle}{j^+} \le 0.9$
	$\left\{-2.88\left(\frac{j_g^+}{j^+}\right)+4.08\right\}\left(1-\sqrt{\frac{\rho_g}{\rho_f}}\right)+\sqrt{\frac{\rho_g}{\rho_f}}$	$V_{gj}^+ = V_{gj,B}^+\exp\left(-1.39\langle j_g^+\rangle\right)$ $+V_{gj,P}^+\left\{1-\exp\left(-1.39\langle j_g^+\rangle\right)\right\}$	Bubbly flow $\langle\alpha\rangle < 0.3$ $0.9 \le \frac{\langle j_g^+\rangle}{j^+} \le 1$
	$1.2\exp\left\{0.110\langle j^+\rangle^{222}\right\}\left(1-\sqrt{\frac{\rho_g}{\rho_f}}\right)+\sqrt{\frac{\rho_g}{\rho_f}}$	$V_{gj}^+ = V_{gj,P}^+$	Cap bubbly flow $\langle\alpha\rangle < 0.3$ $0 \le \langle j^+\rangle \le 1.8$
	$[0.6\exp\{-1.2(\langle j^+\rangle-1.8)\}+1.2]\left(1-\sqrt{\frac{\rho_g}{\rho_f}}\right)+\sqrt{\frac{\rho_g}{\rho_f}}$	$V_{gj}^+ = V_{gj,P}^+$	Cap bubbly flow $\langle\alpha\rangle < 0.3$ $\langle j^+\rangle \ge 1.8$

(continued)

Table 2.2 (continued)

Researchers	Disribution parameter, C_0 (-)	Drift velocity, $\langle\langle v_{gj}\rangle\rangle$ (m/s)	Applicable range
Cunningham and Yeh [7]	$\langle\alpha\rangle = 0.925\left(\frac{\rho_g}{\rho_f}\right)^{0.239}\left[\frac{\langle j_g\rangle}{1.53\left(g\sigma/\rho_f\right)^{1/4}}\right]^{a}\left(\frac{\langle j_f\rangle}{\langle j\rangle}\right)^{0.6}$ where $a = 0.67$ $if\ \frac{\langle j_g\rangle}{1.53\left(g\sigma/\rho_f\right)^{1/4}} < 1$ $a = 0.47$ $if\ \frac{\langle j_g\rangle}{1.53\left(g\sigma/\rho_f\right)^{1/4}} \geq 1$		
Chexal et al. [4]	$L/(K_0 + (1 - K_0)\langle\alpha\rangle^r)$ where $K_0 = B_1 + (1 - B_1)\left(\frac{\rho_g}{\rho_f}\right)^{1/4}$ $L = \min\left[1.15\langle\alpha\rangle^{0.45}, 1.0\right]$ for air-water $r = \frac{1.0 + 1.57\,\rho_g/\rho_f}{1-B_1}$ B_1 is a variable depending on Re. $Re = \begin{cases} Re_g & Re_g > Re_f \ or \ Re_g < 0 \\ Re_f & Re_g \leq Re_f \end{cases}$	$V_{gj}^0 C_g$ where $C_g = (1 - \langle\alpha\rangle)^{B_1}$ $V_{gj}^0 = 1.41\left(\frac{\sigma g \Delta\rho}{\rho_f^2}\right)^{1/4} C_2 C_3 C_4$ C_2, C_3, C_4 are variables depending on ρ_f, $\rho_g, D_H, Re_f = \frac{\langle j_f\rangle D_H \rho_f}{\mu_f}, Re_g = \frac{\langle j_f\rangle D_H \rho_g}{\mu_g}$.	
Shen et al. [45]	1	$\frac{1}{2}\left(\langle\langle v_{gj\_s}\rangle\rangle + \langle\langle v_{gj\_B}\rangle\rangle\right)$ $+\frac{1}{\pi}\left(\langle\langle v_{gj\_s}\rangle\rangle - \langle\langle v_{gj\_B}\rangle\rangle\right) \times arctg\left\{A \times \left(\frac{\langle j_g\rangle}{\langle j_f\rangle + \langle j_g\rangle} - B\right)\right\}$ where $\langle\langle v_{gi\_s}\rangle\rangle = 0.52\sqrt{\frac{Dg\left(\rho_f - \rho_g\right)}{\rho_f}}$ $\langle\langle v_{gi\_B}\rangle\rangle = \sqrt{2}\left(\frac{g\sigma\left(\rho_f - \rho_g\right)}{\rho_f}\right)^{1/4}$ $A = 6.72\left\langle j_f^+\right\rangle^2 - 34.00\left\langle j_f^+\right\rangle + 69.94$ $B = 0.0851\left\langle j_f^+\right\rangle^2 - 0.414\left\langle j_f^+\right\rangle + 0.955$	

correlation. Thus, the applicability of Clark and Flemmer's correlations to other fluid systems such as high pressure steam-water flow is still questionable.

Ishii and Kocamustafaogullari [23] developed a theoretical correlation for the drift velocity for cap bubble flow inside a large diameter channel. Their correlation suggests that the drift velocity increases with the channel diameter and reaches a constant value depending on physical properties at $D_H^* = 30$, corresponding to $D_H = 0.09$ m for air-water systems at atmospheric pressure. The correlation was derived under the assumption that the surface of the cap bubble was smooth. In real two-phase flows, large bubbles can be highly deformed due to natural turbulences in two-phase flow. They did not give any correlation for distribution parameter, however Ishii's correlation for the distribution parameter [20] is usually used with this correlation for the drift velocity.

Hirao et al. [17] conducted experiments with a large pipe of 0.102 m diameter. Their experimental flow conditions included steam-water two-phase flows with liquid and gas volumetric flux below 1 and 4 m/s, respectively. They also proposed a correlation for the drift velocity. It is noted that drift velocity obtained by their correlation can become infinitely large for a very large diameter pipe. This seems not to be physically realizable as well.

Murase et al. [29] performed steam-water two-phase experiments. The test section has four circular channels with an inner diameter 0.0405 and 1.055 m height. A single heater rod is installed inside each flow channel for generating steam flow. The correlations for distribution parameter and drift velocity were developed based on their experimental data.

Kataoka and Ishii [24] developed a drift velocity correlation for pool void fraction based on various experimental conditions. Their correlation shows that the drift velocity is dependent upon vessel diameter, system pressure, gas flux and fluid properties.

Chexal et al. [4] developed a generalized drift-flux correlation for air-water, steam-water, and refrigerant two-phase flows. It depends on pressure, temperature, hydraulic diameter and flow conditions of the two-phase flow but it is independent of flow regime. The correlation covers vertical (up and down), horizontal and inclined cocurrent flows and vertical countercurrent flows. The empirical correlation is usually called EPRI drift-flux model in RELAP 5. The drift-flux correlations for vertical air-water concurrent flows are shown in Table 2.2. The model is entirely empirical, and thus may have some inaccuracies when trying to predict flows in two-phase systems not included in the original benchmarking data set.

Hibiki and Ishii [14] found that the flow regime at the test section inlet affects the liquid circulation pattern resulting in inlet-flow-regime dependent distribution parameter and drift velocity. Two types of drift flux correlations based on different inlet flow regimes, namely bubbly and cap bubbly flows, were proposed. This drift-flux model includes the physical properties of the two phases and relates the distribution parameter and drift velocity to the local liquid and gas flow rates. It should be mentioned that the drift flux correlations developed by Kataoka and Ishii [24] were adopted for cap bubbly flow with void fraction greater than 0.3 in their drift flux model. In Table 2.2, $V_{gj,P}$ is the pool-boiling drift velocity developed by Kataoka and Ishii [24], while $V_{gj,B}$ is the bubbly flow drift velocity given by Ishii [20].

Shen et al. [45] found an *N*-shaped void fraction development (increase-decrease-increase) of the void fraction in the flow direction as the two-phase flow develops from bubbly flow from to slug (large cap) flow at low liquid flow rates and developed a drift flux model for the two-phase flow ranging from bubbly flow from to slug flow in a vertical large diameter pipe to predict the *N*-shaped changing void fraction.

Other than the various drift-flux models, Cunningham and Yeh [7] developed a void fraction prediction model for rod bundle systems under pool conditions based on data from a simulated small-break LOCA (Loss-Of-Coolant Accident) experiment in a PWR fuel bundle assembly (480 heated rods). The model is a practical empirical correlation, but without the general form of the drift-flux model using a distribution parameter and a drift velocity.

2.5 Void Fraction Data in Large Diameter Pipes

Void fraction prediction in large diameter channels is essential for many applications. Because of this, many studies have been performed to measure the void fraction in large diameter pipes. These studies have been listed in Table 2.3 and are summarized here.

Among the first of these tests were those performed by Baily et al. [1] who made void fraction measurements in pipes ranging from 0.152 to 0.610 m in diameter using an air-water system. This experiment was able to provide valuable data on the bubbly to cap-turbulent transition region, but was limited to void fractions smaller than 0.4 and to pool conditions. This means that the data does not provide any information on the behavior of churn-turbulent flows or flows with higher liquid fluxes.

Later several high-pressure steam-water flow experiments were performed to study the behavior of nuclear reactor systems during accident scenarios [3, 53, 56]. These studies provide important data for higher void fraction flows, but again are limited to pool conditions.

Several further experiments were performed by Hills [16] in a 0.152 m diameter pipe, but void fractions were limited to 0.55. Important data regarding low void fraction behavior at varied liquid velocities was obtained, but data at high void fractions that would allow evaluation of the churn-turbulent flow regime was not. Hall et al. [10] performed steam-water experiments in a 0.171 m diameter pipe for low liquid velocities and void fractions of up to 0.9. This data is valuable for evaluating the transition to annular flow, however very few data points were collected in this region. Hirao et al. [17] measured the void fraction in steam-water two-phase flow using a large scale apparatus with 0.1023 m in diameter. The data were obtained in a large range of flow conditions, but the number of data points is

Table 2.3 Void fraction data sets for large diameter channels

Researchers	Pipe diameter, D (m)	Superficial liquid velocity, $\langle j_f \rangle$ (m/s)	Superficial gas velocity, $\langle j_g \rangle$ (m/s)	Void fraction, $\langle a \rangle$ (-)	Pressure, P (MPa)	Fluid system	Number of data points
Baily et al. [1]	0.152, 0.304, 0.610	0	up to 0.53	up to 0.4	~0.1	Air-Water	47
Wilson et al. [56]	0.101, 0.480	0	up to 1.9	up to 0.85	1.0–4.1	Steam-Water	115
Carrier [3]	0.456	0	up to 1.2	up to 0.82	4.1–13.8	Steam-Water	53
Styrikovich and Kutateladze [53]	0.238	0	up to 1.6	up to 0.7	0.61–9.3	Steam-Water	57
Hills [16]	0.150	up to 2.6	up to 3.2	up to 0.55	~0.1	Air-Water	394
Hall et al. [10]	0.171	up to 0.2	up to 0.75	up to 0.9	4.4	Steam-Water	90
Hashemi et al. [11]	0.305	up to 0.06	up to 1.2	up to 0.94	~0.1	Air-Water	30
Ohnuki and Akimoto [30]	0.480	up to 0.2	up to 0.9	up to 0.5	~0.1	Air-Water	75
Ohnuki and Akimoto [31]	0.200	up to 1	up to 4.7	up to 0.55	~0.1	Air-Water	58
Hirao et al. [17]	0.1023	up to 1.0	up to 4.0	up to 0.7	~1.5	Steam-Water	25
Hibiki and Ishii [13]	0.101	up to 0.4	up to 0.3	up to 0.3	~0.1	Nitrogen-Water	467
Yoneda et al. [57]	0.152	up to 0.6	up to 0.25	up to 0.35	~0.5	Steam-Water	12
Inoue [19]	0.200	up to 0.3	up to 0.35	up to 0.47	~0.1	Air-Water	84
Smith [50]	0.101, 0.152	up to 2	up to 8	up to 0.71	~0.1	Air-Water	106
Sawant et al. [36]	0.152	up to 2	up to 5.2	up to 0.83	~0.1	Air-Water	248
Shawkat et al. [42]	0.200	up to 0.7	up to 0.18	up to 0.15	~0.1	Air-Water	34
Schlegel et al. [37]	0.152, 0.203	up to 1	up to 8	up to 0.8	0.180–0.280	Air-Water	655
Prasser [60]	0.194	1 m/s	up to 0.53	up to 0.3	6.5	Steam-Water	7
Omebere-Iyari et al. [34]	0.194	up to 0.65	up to 0.95	up to 0.58	~4.64	Steam-Water	10
Shen et al. [45]	0.200	up to 0.311	up to 0.510	up to 0.39	~0.1	Air-Water	128

small. Hashemi et al. [11] performed experiments in a 0.305 m diameter facility with void fractions as high as 0.94, but again only very few data points were collected at such high void fractions.

A series of experiments investigating flow development in large diameter pipe systems were performed by Ohnuki and Akimoto [30, 31]. These experiments provided valuable insight into flow structure, but were again limited to void fractions less than 0.5. Yoneda et al. [57] performed some experiments using steam-water flows in pipes with diameters of up to 0.152 m, however few tests were reported. Also Inoue [19] performed air-water tests in a facility with 0.2 m diameter and for void fractions up to 0.47.

In relation to nuclear reactor accident and safety studies, Hibiki and Ishii [13] performed experiments focusing on bubbly flow behavior in a 0.101 m diameter pipe to understand the two-phase natural circulation and flow termination during a small break loss of coolant accident in light water reactors. One of the important aspects in their experiments was that inlet flow conditions were varied by inserting or removing a horizontal header section between the gas injector and the hot leg in order to investigate the effect of the gas phase inlet conditions on the flow regimes and flow interruption.

Sawant et al. [36] performed experiments in a 0.152 m diameter pipe with void fractions of up to 0.83, providing valuable data on flow pattern and flow behavior in the cap-turbulent and churn-turbulent regions. Shawkat et al. [42] performed very low void fraction experiments in a 0.2 m diameter pipe to evaluate the void profile, gas and liquid velocity profiles. Schlegel et al. [37] performed extensive experiments in pipes with 0.152 and 0.203 m diameter with void fractions as high as 0.8 to provide valuable data on the transition from cap-turbulent to churn-turbulent flows.

Omebere-Iyari et al. [34] carried out measurements of steam/water two-phase flow by using a wire mesh sensor in the 0.194 m diameter vertical pipe at the pressure of 4.64 MPa. Their data were taken in churn turbulent flow with void fraction as high as 0.58.

Shen et al. [45] performed experiments in a 0.2 m diameter pipe with a total length of 26 m at low liquid flow rates. They found that the void fraction tends to increase in the inlet region due to expansion of small spherical bubbles. This is followed by a decrease in the void fraction along the flow channel caused by an increase in the relative velocity as small bubbles begin to coalesce into larger Taylor bubbles. Eventually the void fraction begins to increase again as the coalescence rate decreases and expansion of Taylor bubbles begins to become important. This axial void fraction profile is shown in Fig. 2.16, which illustrates the evolution of the flow from dispersed, spherical bubbles to cap-bubbly flow in large diameter pipes.

In addition, early work in large diameter pipes was performed by Filiminov et al. [8], Styrikovich et al. [52] and Petukhov and Kolokoltsen [35] for steam-water flow in larger diameter channels. These studies have been published in Russian and the results were utilized in the development of Kataoka and Ishii's [24] correlation.

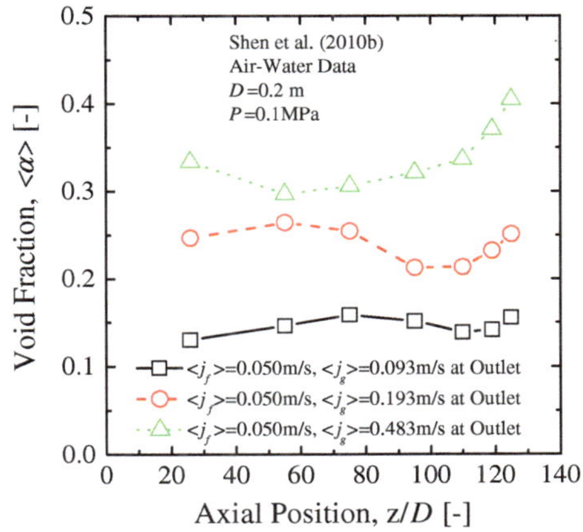

Fig. 2.16 *N*-shaped void fraction change along the flow development at low flow rate

These databases widely cover extensive experimental conditions such as flow channel diameters (0.102–0.480 m), pressures (0.1–13.8 MPa), superficial liquid velocity (0.0–2.6 m/s), superficial gas velocity (0.0–8.0 m/s), void fraction (0–0.94), bubble injection methods (bubbly or cap-bubbly injection), and fluid systems (air-water, nitrogen-water, and steam-water). However, the available data for void fraction in large diameter pipes still has some significant limitations. Especially for forced convection flows, data for the transition to annular flow region is insufficient to validate the models or flow regime transitions.

2.6 Comparisons and Analysis

In this section, the various models described in the previous section are evaluated based on the existing database of two-phase flow measurements in large diameter pipes.

2.6.1 Drift Velocity in Large Diameter Pipes

The drift velocity in the drift-flux model is the result of the relative velocity between the gas and liquid phases. Thus the increased relative velocity of bubbles in large diameter channels is expressed in the drift-flux model using the drift velocity. For relatively low void fractions such as those encountered in bubbly flow the drift velocity is determined by the rise velocity of small, spherical bubbles. Thus under these conditions the drift velocity in large diameter channels is similar to that in small diameter channels. When the flow condition becomes cap-turbulent however, the larger relative velocity of the cap bubbles results in a significantly higher drift

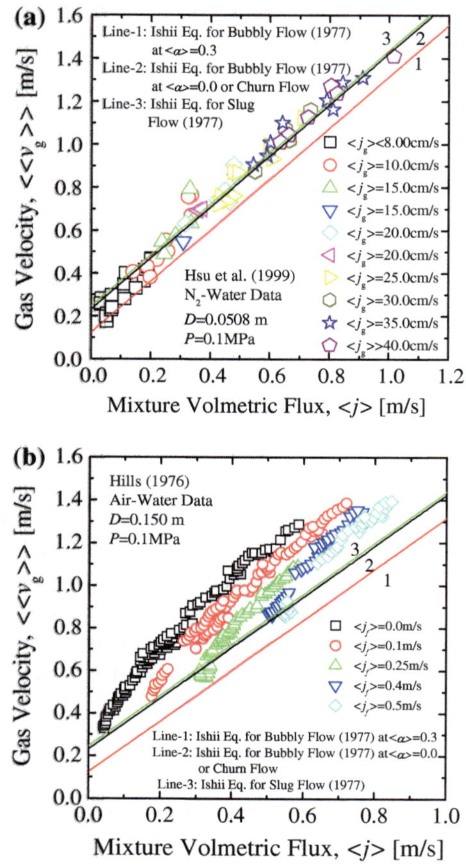

Fig. 2.17 Comparison of small diameter pipe data (D = 50.8 mm) with large diameter pipe data (D = 150 mm) at P = 0.1 MPa. **a** Hsu et al. [18] data for a small diameter pipe with D = 50.8 mm. **b** Hills [16] data for a large diameter pipe with D = 150 mm

velocity for drift-flux models in large diameter channels [14]. To illustrate this effect, data from a pipe with non-dimensional diameter of 18.5 (D = 50.8 mm, [18]) and 55 (D = 152 mm, [16]) are compared with the drift-flux correlations developed by Ishii [20] for small pipes in Fig. 2.17.

As shown in Fig. 2.17a, the data for bubbly flow in a vertical pipe with a non-dimensional diameter of 18.5 collapse to a single line. The drift flux correlations of Ishii [20] with the drift velocity for bubbly or churn or slug flow can predict the void fraction in the small diameter pipe over all flow conditions tested in the experiment of Hsu et al. [18]. On the other hand, Fig. 2.17b reveals the effect of the pipe diameter on the ability of drift flux correlations to predict the void fraction. The data for bubbly flow in a vertical pipe with an inner diameter of 150 mm do not fall on a single line. Generally the drift velocity gradually increases from that for bubbly flow to that for cap-bubbly flow over a transition region between bubbly flow and cap-bubbly flow as large bubbles begin to occur and eventually become dominant.

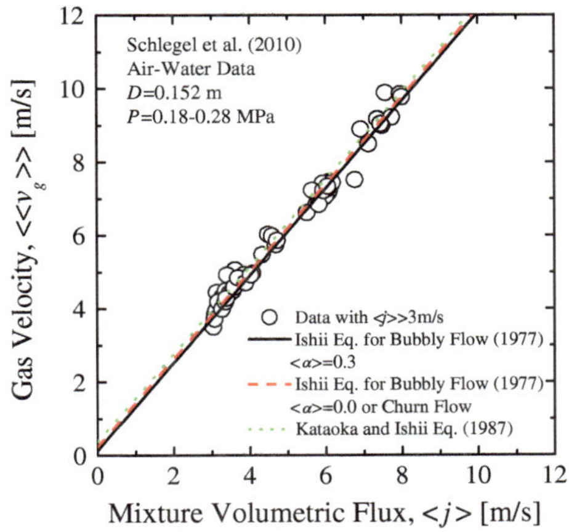

Fig. 2.18 Comparison of void fraction measured in a large pipe ($D = 0.152$ m) at high mixture volumetric flux with various drift flux correlations

2.6.2 Distribution Parameter in Large Diameter Pipes

Within the flow channel, there exists a profile of both velocity and void fraction. The result of these combined effects is that the gas phase tends to concentrate in the higher-velocity region in the center of the pipe [2]. This means that the gas is concentrated in a region of high velocity, while the liquid is concentrated in a region of low velocity. This profile effect is accounted for in the drift-flux model through the distribution parameter.

At low total volumetric fluxes, the change in drift velocity alone is not sufficient to explain all of the variation shown in Fig. 2.17. In addition, the discussion in Sect. 2.2.5 shows that the void and velocity profiles can vary significantly from those typically seen in small diameter pipes. This leads to significant changes in the distribution parameter under these conditions [14].

As the total mixture velocity is increased, the importance of the relative velocity and therefore the drift velocity decreases, while the effect of the void profile, expressed using the distribution parameter, becomes dominant. At very high mixture volumetric flux, it is not unreasonable to neglect the relative velocity and consider only the profile effect.

To illustrate this, the void fraction data collected by Schlegel et al. [37] has been compared to various drift-flux correlations in Fig. 2.18. The predicted drift velocities from Ishii's bubbly and churn flow equations and Kataoaka and Ishii equation are 0.124, 0.231 and 0.381 m/s respectively for the experimental conditions shown in the figure. The difference in these drift velocity predictions does not significantly affect the prediction accuracy of the drift-flux model for these high mixture volumetric flux conditions. Figure 2.18 indicates that the distribution parameter given by Ishii [20] can be applicable to upward two-phase flow in a large diameter pipe at high mixture volumetric flux conditions.

2.6.3 Drift Flux Type Correlation Evaluations

In view of the great difference in flow characteristics for various flow regimes, the existing drift flux models are evaluated in two steps: for low and high void fraction flows respectively. The low void fraction flows correspond to the bubbly and bubbly-slug transition regions and the high void fraction flows correspond to the cap turbulent and churn turbulent flow regimes. The evaluation of these data sets with regards to the various models described previously is shown in the following figures. The first several figures, Figs. 2.19, 2.20, 2.21, 2.22, 2.23, 2.24, 2.25, 2.26, 2.27, 2.28, 2.29 and 2.30, show the comparison of bubbly flow data with the various models for void fractions smaller than 0.3. The remaining figures, Figs. 2.31, 2.32, 2.33, 2.34, 2.35, 2.36, 2.37, 2.38, 2.39 and 2.40, show the comparison of higher void fraction data, for void fractions larger than 0.3, with various models.

2.6.3.1 Evaluations with Low Flow Rate and Low Void Fraction Data

Figure 2.19 shows the comparison of data collected in a 0.102 m diameter pipe with various drift-flux correlations as a parameter of superficial gas velocity. Figure 2.19a shows the behavior of bubbly flows at low void fraction when the inlet flow regime is uniformly distributed bubbly flow. The bubbly data do not fall on a single line at low mixture volumetric flux. As the figure shows none of the correlations shown can accurately predict all of the data for bubbly flows in large diameter systems. Figure 2.19b shows the behavior of the flow when it is injected as cap-bubbly flow for void fractions smaller than 0.3, showing the strong effect that the inlet condition has on the flow. In this case, the flow behavior is mainly determined by the behavior of cap bubbles, which is why the data is predicted well by only a single line.

Figure 2.20 shows the comparison of Hills [16] correlation with the data of Hibiki and Ishii [13]. The overall agreement is fair; however the correlation cannot reproduce the behavior at very low void fraction and may not perform very well at higher liquid fluxes. Figures 2.21 and 2.22 show the comparison of Hibiki and Ishii's [13] data with the predictions using correlations of Shipley [48] and Clark and Flemmer [5] respectively. Neither model shows a reasonable agreement with the data. Figure 2.23 compares the correlation of Clark and Flemmer [6] with the data of Hibiki and Ishii [13]. The overall agreement between them is fairly good. However, for $\left\langle j_g^+ \right\rangle < 0.4$ the correlation overestimates the gas velocity, whereas for $0.5 < \left\langle j_g^+ \right\rangle < 1.2$ the correlation underestimates the gas velocity. It should be pointed out that these correlations were developed based on adiabatic air-water flow data. Since physical properties would affect the drift velocity significantly [24], the applicability of these correlations to high pressure or other fluid systems has not been validated.

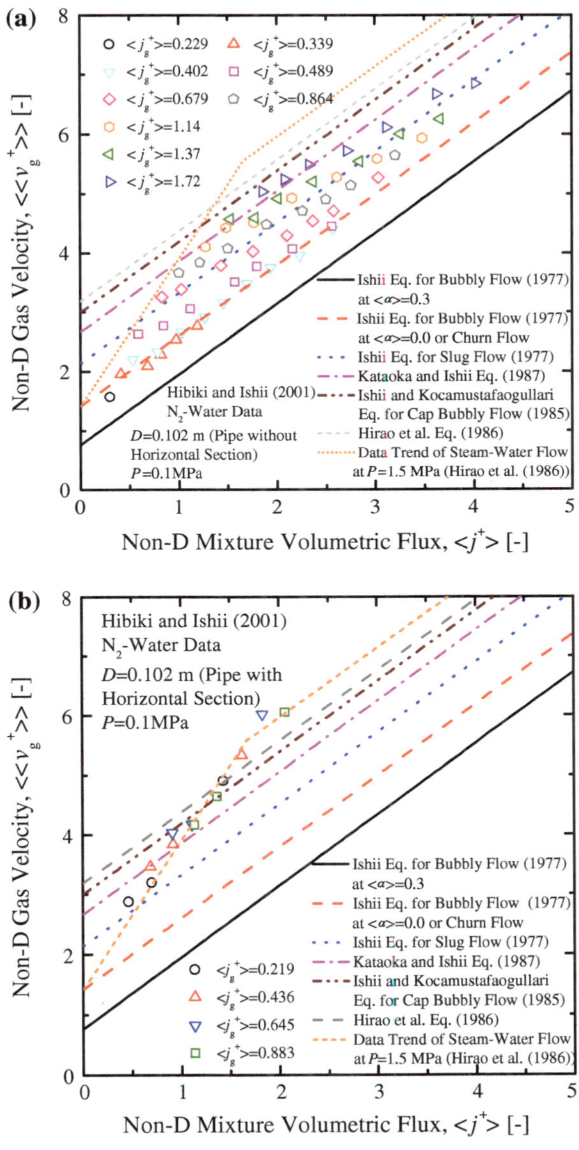

Fig. 2.19 Comparison of various drift-flux correlations with air-water data in 0.102 m diameter pipe [13]. **a** The inlet flow regime is bubbly flow. **b** The inlet flow regime is cap-bubbly flow

The correlation of Murase et al. [29] is compared with the data of Hibiki and Ishii [13] in Fig. 2.24. The correlation does not predict the gas velocity well. The comparison with the correlation of Cunningham and Yeh [7] is shown in Fig. 2.25. Although the correlation was developed from data collecting during core uncovering experiments (namely pooling conditions), it greatly overestimates the low void fraction and low liquid flow rate data of Hibiki and Ishii [13].

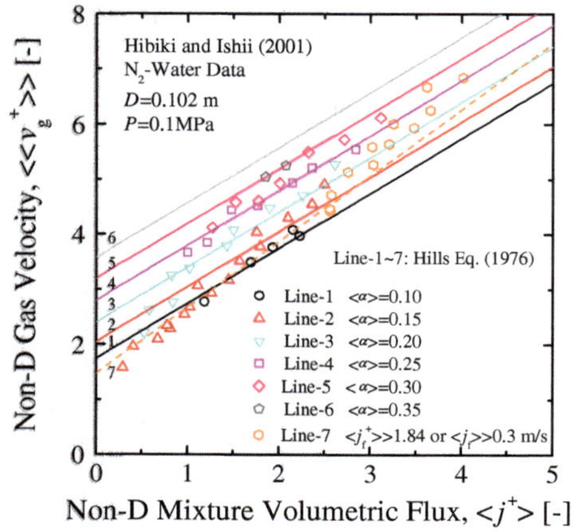

Fig. 2.20 Comparison of Hills correlation [16] with air-water data in 0.102 m diameter pipe [13]

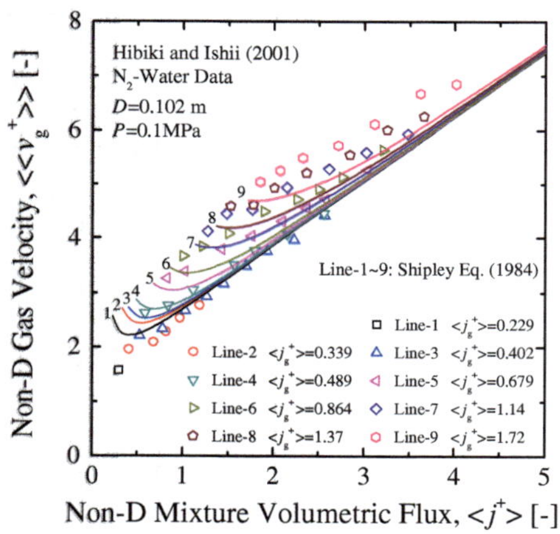

Fig. 2.21 Comparison of Shipley correlation [48] with air-water data in 0.102 m diameter pipe [13]

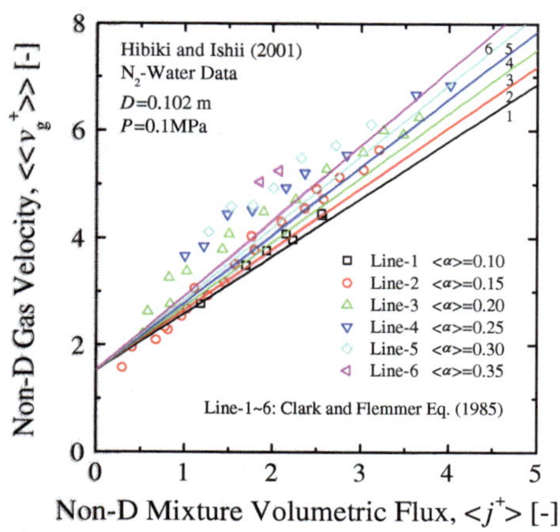

Fig. 2.22 Comparison of Clark and Flemmer correlation [5] with air-water data in 0.102 m diameter pipe [13]

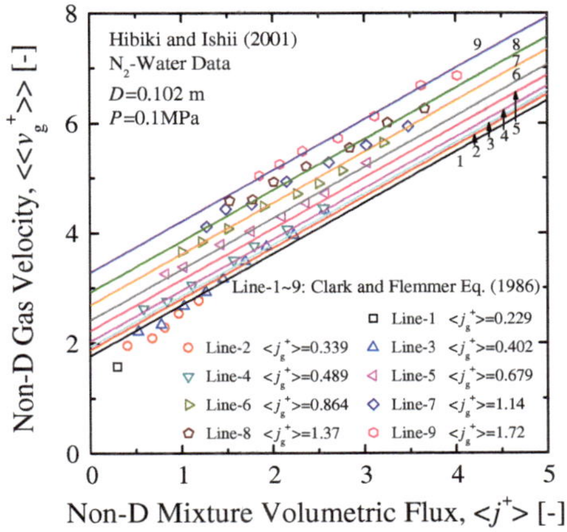

Fig. 2.23 Comparison of Clark and Flemmer correlation [6] with air-water data in 0.102 m diameter pipe [13]

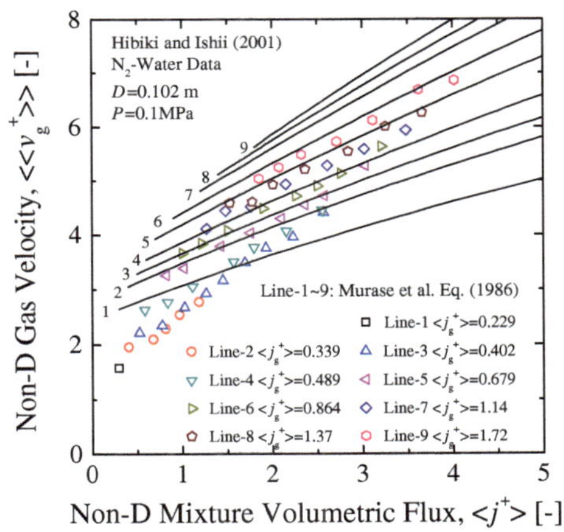

Fig. 2.24 Comparison of Murase et al. correlation [29] with air-water data in 0.102 m diameter pipe for bubbly flow injection [13]

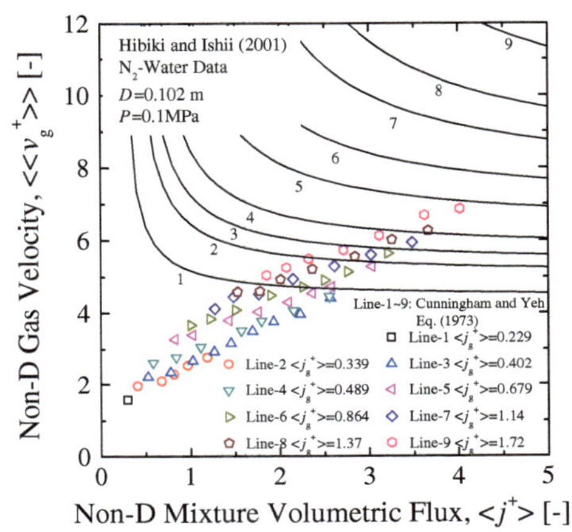

Fig. 2.25 Comparison of Cunninghame and Yeh correlation [7] with air-water data in 0.102 m diameter pipe for bubbly flow injection [13]

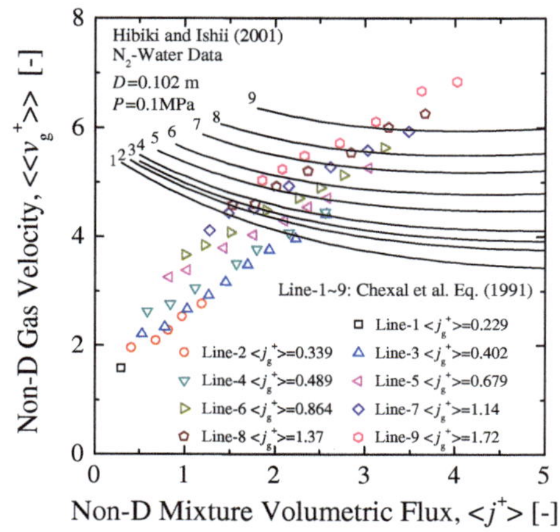

Fig. 2.26 Comparison of Chexal et al. correlation [4] with air-water data in 0.102 m diameter pipe for bubbly flow injection [13]

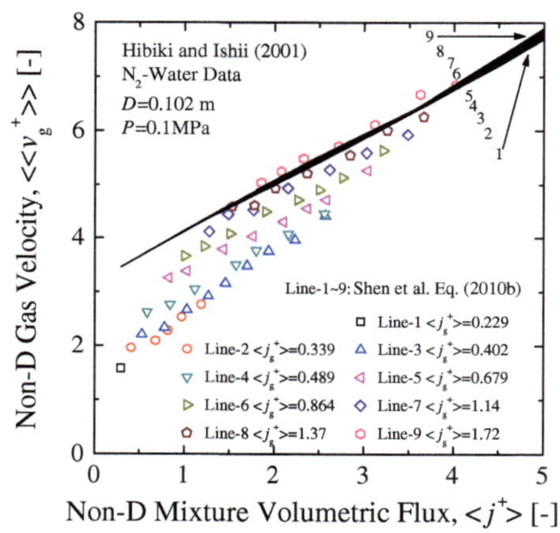

Fig. 2.27 Comparison of Shen et al. correlation [45] with air-water data in 0.102 m diameter pipe for bubbly flow injection [13]

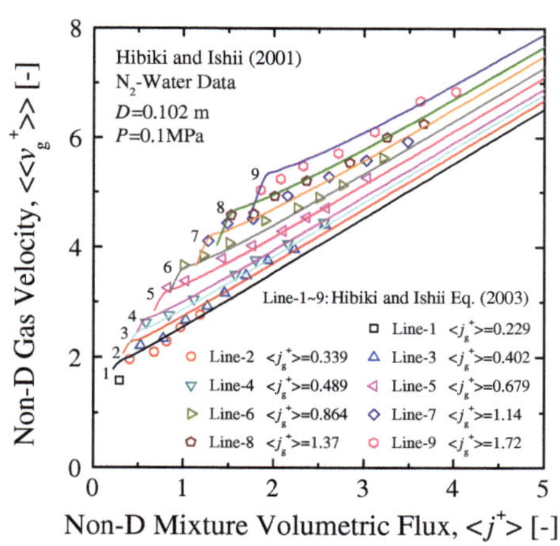

Fig. 2.28 Comparison of Hibiki and Ishii correlation [14] with air-water data in 0.102 m diameter pipe for bubbly flow injection [13]

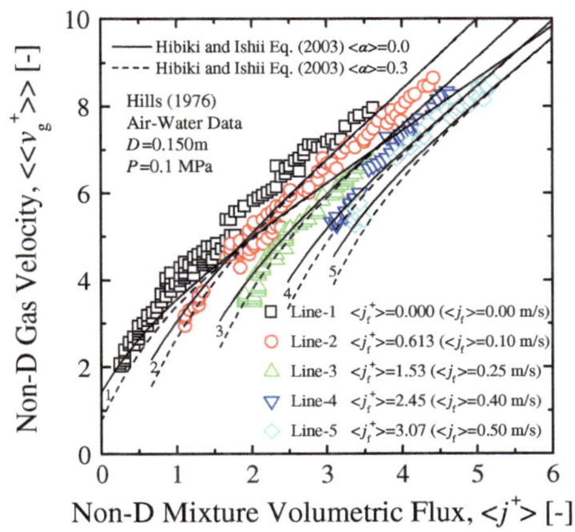

Fig. 2.29 Comparison of Hibiki and Ishii correlation [14] with air-water data taken by Hills [16] in 0.150 m diameter pipe

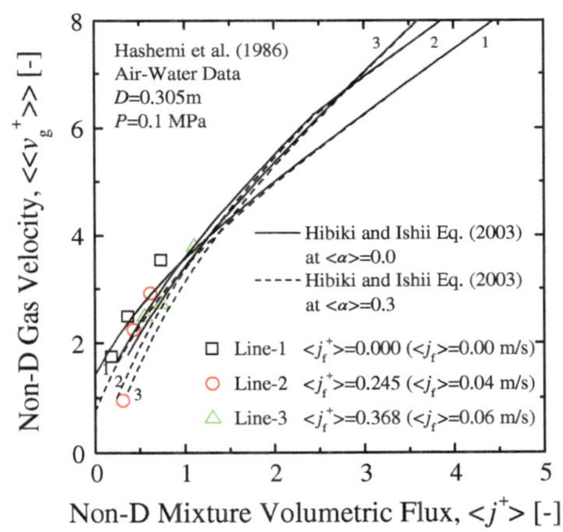

Fig. 2.30 Comparison of Hibiki and Ishii correlation [14] with air-water data of Hashemi et al. [11] taken in 0.305 m diameter pipe

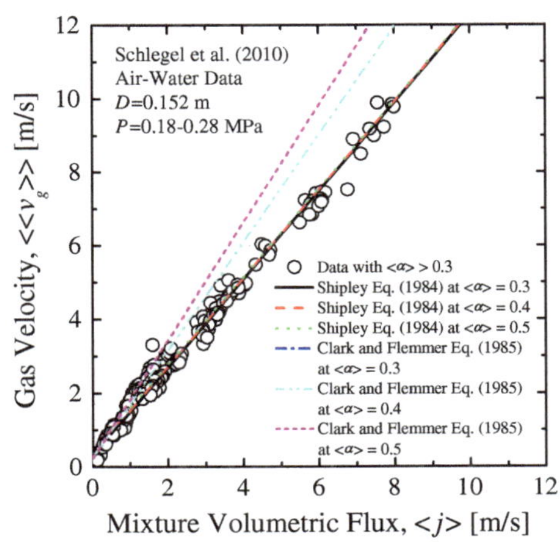

Fig. 2.31 Comparison of the correlations of Shipley [48] and Clark and Flemmer [5] with the data of Schlegel et al. [37]

Fig. 2.32 Comparison of the correlations of Ishii [20], Hirao et al. [17], and Clark and Flemmer [6] with the data of Schlegel et al. [37]

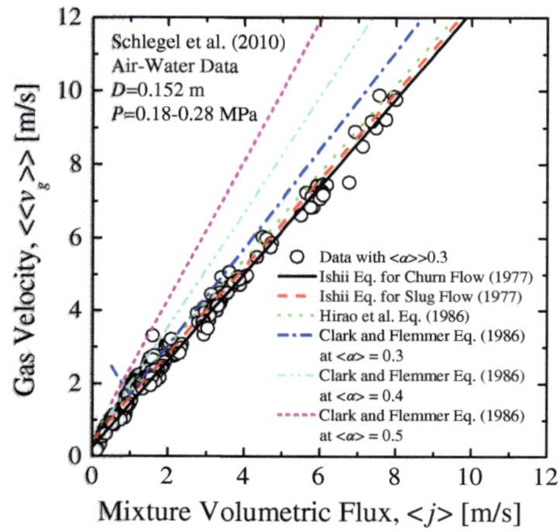

Fig. 2.33 Comparison of the correlations of Kataoka and Ishii [24] and Hills [16] with the data of Schlegel et al. [37]

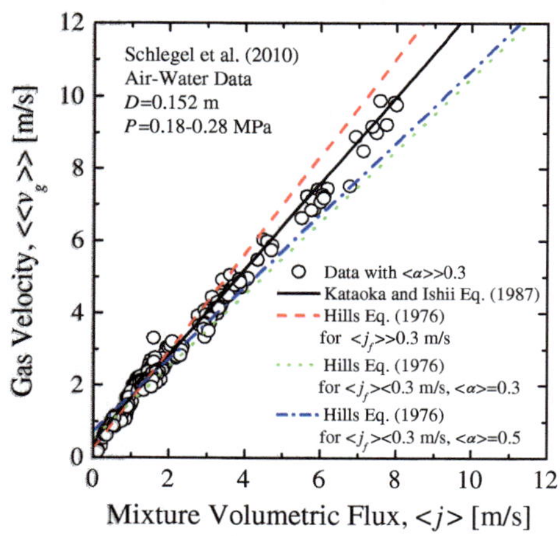

Fig. 2.34 Comparison of the correlations of Cunningham and Yeh [7] and Murase et al. [29] with the data of Schlegel et al. [37]

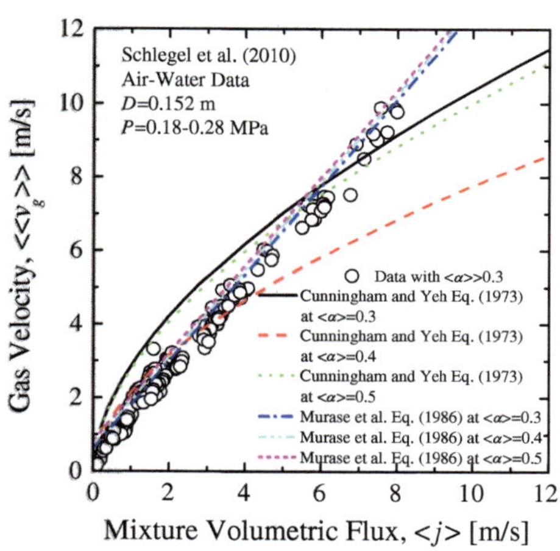

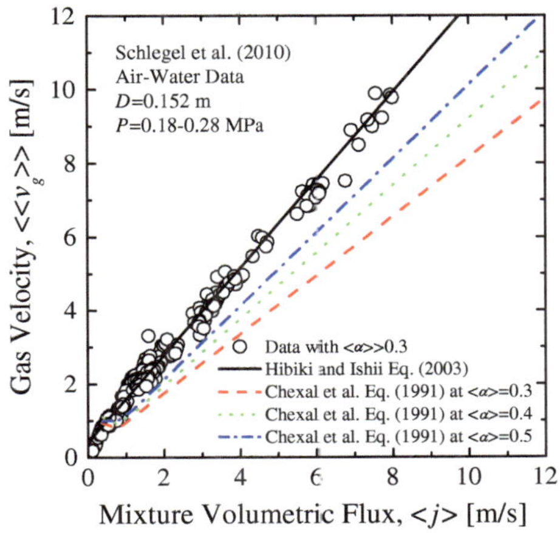

Fig. 2.35 Comparison of the correlations of Hibiki and Ishii [14] and Chexal et al. [4] with the data of Schlegel et al. [37]

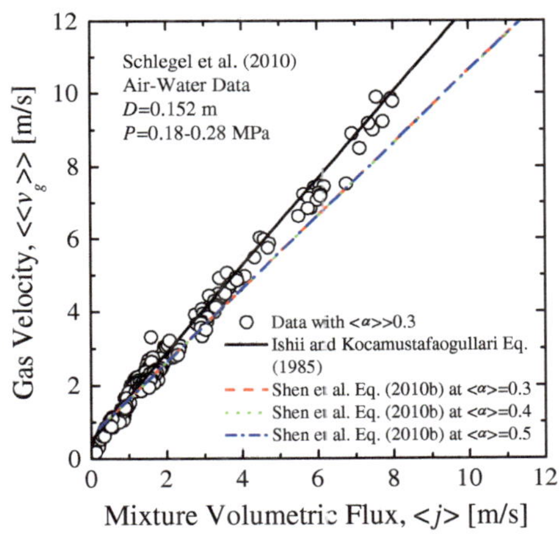

Fig. 2.36 Comparison of the correlations of Ishii and Kocamustafaogullari [23] and Shen et al. [45] with the data of Schlegel et al. [37]

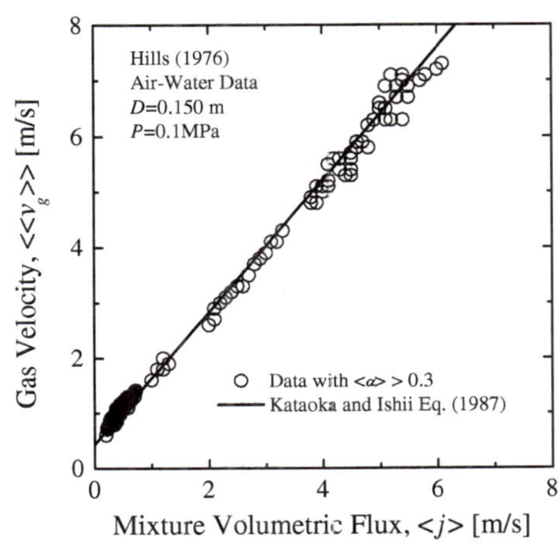

Fig. 2.37 Comparison of Hills' data [16] with the drift-flux model of Kataoka and Ishii [24] at high void fraction

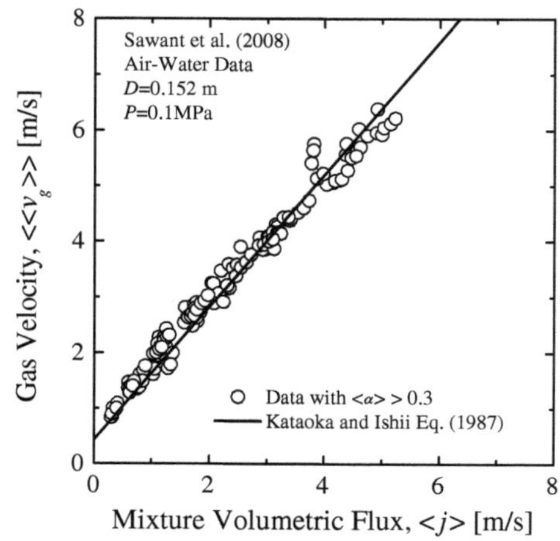

Fig. 2.38 Comparison of Sawant et al. [36] data with the drift-flux model of Kataoka and Ishii [24] at high void fraction

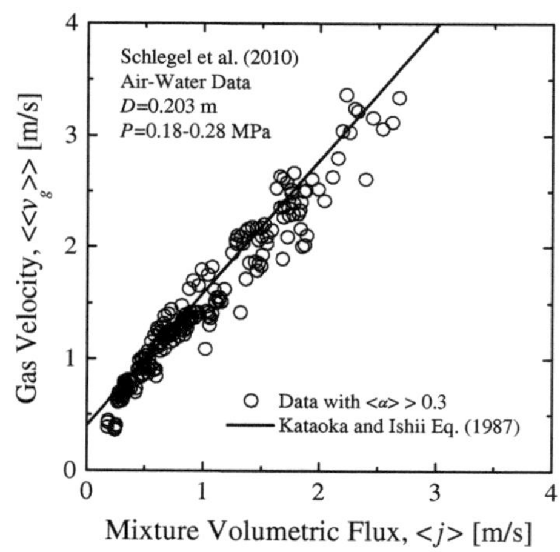

Fig. 2.39 Comparison of Schlegel et al. [37] data with the drift-flux model of Kataoka and Ishii [24] at high void fraction

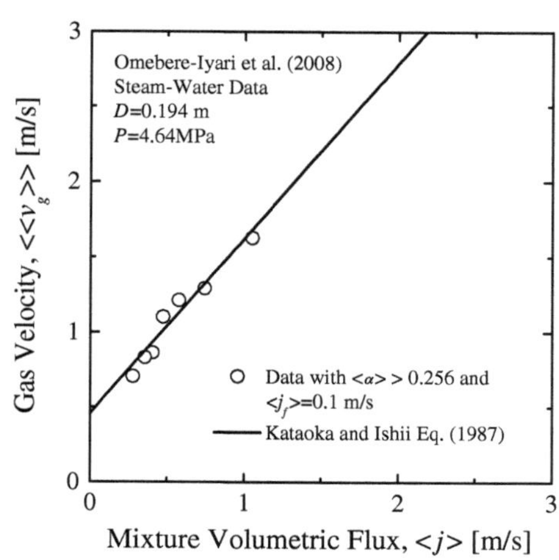

Fig. 2.40 Comparison of Omebere-Iyari et al. data [34] with the drift-flux model of Kataoka and Ishii [24] at high void fraction

In Fig. 2.26, the correlation developed by Chexal et al. [4] is presented in comparison with the data of Hibiki and Ishii [13]. The correlation of Chexal et al. [4] was developed based on a wide range data sets and has been shown to effectively predict the void fraction of the data set on which it is based. Figure 2.26 shows that its prediction for the experimental conditions of Hibiki and Ishii [13] does not capture the change in gas velocity as the mixture volumetric flux changes and the overall accuracy is not very good.

The correlation of Shen et al. [45] is compared with the data of Hibiki and Ishii [13] in Fig. 2.27. Although the correlation can predict the data in the flow transition from bubbly to cap turbulent flows, its agreement with the data of Hibiki and Ishii [13] is not acceptable and it's accuracy at higher total flux conditions is questionable.

The comparison of the correlation developed by Hibiki and Ishii [14] with the data collected earlier by the same authors [13] is shown in Fig. 2.28. The solid lines indicate the predictions of the model for each superficial gas velocity. This drift-flux model, though complex, shows that it can accurately reproduce the experimentally measured gas velocity trends and values very well in Fig. 2.28.

The comparisons of the experimental data taken by Hibiki and Ishii [13] with the drift-flux type correlations in Figs. 2.19, 2.20, 2.21, 2.22, 2.23, 2.24, 2.25, 2.26, 2.27 and 2.28 show that the correlations developed by Hibiki and Ishii [14] gives the most reasonable predictions of the experimental data in large diameter pipes at low liquid flow rate and low void fraction conditions.

The correlation developed by Hibiki and Ishii [14] has also been compared to the data of other researchers. Figures 2.29 and 2.30 shows the comparisons of the correlations with the experimental data taken by Hills [16] and Hashemi et al. [11] at low liquid flow rate and low void fraction conditions. Solid and broken lines in the Figs. 2.29 and 2.30 indicate calculated values by the correlation of Hibiki and Ishii [14] at $\langle\alpha\rangle = 0$ and 0.3, respectively. Since Hills [16] and Hashemi et al. [11] carried out the experiments by changing the superficial gas velocity, keeping the superficial liquid velocity constant, the data in Figs. 2.29 and 2.30 are presented as a parameter of the superficial liquid velocity. The drift-flux correlation of Hibiki and Ishii [14] can predict the proper trend and the value of the experimental data very well. The correlation is found to have an average relative prediction error of ± 6.7 % for the 373 data sets of Hibiki and Ishii [13], Hills [16] and Hashemi et al. [11] shown in Figs. 2.28, 2.29, 2.30.

2.6.3.2 Evaluations with High Void Fraction Data

When the void fraction is higher than 0.3, the large cap bubbles are formed and play dominant roles in determining the flow behavior in large diameter pipes. The corresponding flow regimes fall into the regions of cap turbulent flow and churn turbulent flow. The flow characteristics in the high void fraction regions are expected to be different from those in the low void fraction regions. The data taken by Schlegel et al. [37] in pipes with 0.152 m diameter covered an extensive range of void fractions. Among these data, the data with high void fraction ranging from 0.3 to 0.8 are valuable and typical. They are selected to evaluate all of the available drift-flux correlations in the flow regime transition from cap turbulent to churn turbulent flows.

Figure 2.31 shows the comparison of the high void fraction data collected in a 0.152 m diameter pipe with the drift-flux correlations of Shipley [48] and Clark and Flemmer [5]. Both of the correlations can predict the data at low mixture volumetric flux. Since Clark and Flemmer [5] did not give a physically accurate prediction for the distribution parameter the model agreement with high mixture volumetric flux data is not good. The correlations of Shipley [48] can predict the whole data well.

Comparison of the correlations of Ishii [20], Hirao et al. [17] and Clark and Flemmer [6] with the data of Schlegel et al. [37] is presented in Fig. 2.32. The churn and slug flow correlations of Ishii [20] were originally developed for high void fraction flow in a small diameter pipe. Although these models can acceptably predict the data for high mixture volumetric flux in large diameter pipes, this is due to a hydrodynamically accurate model for the distribution parameter. At lower mixture volumetric fluxes these models underestimate the data of Schlegel et al. [37] due to the differences in the drift velocity. The correlation of Clark and Flemmer [6] overestimated the data. The correlation of Hirao et al. [17] gave a reasonable prediction for the data.

The correlations of Kataoka and Ishii [24] and Hills [16] are compared with the data of Schlegel et al. [37] in Fig. 2.33. The correlation of Hills [16] predicts the high void fraction data well in the mixture volumetric flux ranging from 0.2 to 3 m/s. For low liquid flow rates however, the correlation underestimates the gas velocity and at high gas flow rates or pool conditions the correlation overestimates the gas velocity. The correlation of Kataoka and Ishii [24] was developed for the cap turbulent flow in pool conditions for large diameter pipes. Because of this, it reasonably predicts the high void fraction data of Schlegel et al. [37].

Figure 2.34 depicts the comparison of the correlations of Cunningham and Yeh [7] and Murase et al. [29] with the data of Schlegel et al. [37]. The correlation of Cunningham and Yeh [7] did not reasonably predict the data and its void fraction estimation is questionable in cap turbulent flow in a large diameter pipe. The correlation of Murase et al. [29] acceptably predicted the data of Schlegel et al. [37].

Comparison of the correlations of Hibiki and Ishii [14] and Chexal et al. [4] with the data of Schlegel et al. [37] is illustrated in Fig. 2.35. The generalized drift flux model of Chexal et al. [4] did not perform well. In most regions, it greatly underestimates the gas velocities in the large diameter pipe. Since the correlation of Hibiki and Ishii [14] is identical to the correlation of Kataoka and Ishii [24] in this region, it satisfactorily predicted the high void fraction data of Schlegel et al. [37].

Figure 2.36 shows the comparison of the correlations of Ishii and Kocamustafaogullari [23] and Shen et al. [45] with the data of Schlegel et al. [37]. Since the distribution parameter of Shen et al. [45] is set to be 1, its prediction for gas velocities in the high mixture volumetric flux region is not good. However, the drift velocity correlation of Shen et al. [45] is similar to the correlation of Hirao et al. [17] for these experimental conditions and can predict the void fraction at low mixture volumetric flux. The prediction of Ishii and Kocamustafaogullari [23] shows reasonable agreement with the experimental data of Schlegel et al. [37].

From the above-mentioned discussions on the comparisons of the available drift flux models with the high void fraction data of Schlegel et al. [37] in Figs. 2.31, 2.32, 2.33, 2.34, 2.35 and 2.36, we know that the correlations of Shipley [48], Hirao et al. [17], Ishii and Kocamustafaogullari [23], Kataoka and Ishii [24] and Hibiki and Ishii [14] can satisfactorily predict the data in a large diameter pipe. Since physical properties are not included in the correlation of Shipley [48], it may be questionable to apply the correlation to high pressure or other fluid systems. The dependence of Hirao et al. [17] on diameter may not be physically realistic at very high diameters and therefore this correlation should not be used. Hibiki and Ishii [14] adopt the correlation of Kataoka and Ishii [24] when void fraction is greater than 0.3. So the drift-flux correlation of Kataoka and Ishii [24] is recommended for the high void fraction calculation in a large diameter pipe.

Many data sets for large diameter pipes other than that of Schlegel et al. [37] are available for relatively high void fraction conditions. Data was collected in a 0.152 m diameter pipe at 0.1 MPa pressure using an air-water mixture by Hills [16] and Sawant et al. [36]. Additional data was collected by Schlegel et al. [37] in a pipe with diameter of 0.203 m. Omebere-Iyari et al. [34] collected data using a steam-water system at pressure of 4.64 MPa in a pipe with diameter of 0.194 m. This data has been compared to the correlation of Kataoka and Ishii [24] in Figs. 2.37, 2.38, 2.39 and 2.40.

The comparison of the drift-flux correlation of Kataoka and Ishii [24] with these high void fraction data ranging from 0.1 to 4.64 MPa pressure and from air-water to steam-water shows good agreement. This serves to further illustrate the effectiveness of the drift-flux correlation of Kataoka and Ishii [24] for high void fraction flows. The impacts of different fluids and different pressures are included in the drift-flux correlation of Kataoka and Ishii [24] through the physical properties.

The comparisons in Figs. 2.33, 2.37, 2.38, 2.39 and 2.40 also illustrated that the pipe diameter ranging from 0.15 to 0.203 m has little effect on the behavior of the cap turbulent flow and churn turbulent flow and the performance of the drift-flux models in large diameter pipes.

In view of the evaluations of the 13 drift-flux type correlations from low and high void fraction data available now, the drift-flux correlations of Hibiki and Ishii [14] and Kataoka and Ishii [24] perform satisfactorily for low and high void fraction conditions respectively. They are recommended for void fraction prediction in the two-phase flow in large diameter pipes. Since the correlations for the distribution parameter and the drift velocity in a large diameter pipe have not been validated separately by detailed local flow data, they should not be used individually. In a future study, detailed local measurements of flow parameters for gas and liquid phases in a large diameter pipe are recommended to develop a detailed and more rigorous drift-flux correlation taking account of the detailed flow structure.

Because of the importance of predicting two-phase flows in large diameter pipes, many studies have been performed to measure the void fraction and other two-phase flow characteristics in such systems. These experiments have provided valuable insight into the behavior and structure of the flow, however there are still some limitations to the existing database. Very high void fraction data in the churn-turbulent and annular/mist flow regimes is still very limited, which inhibits the ability of researchers to develop models to predict the flow in this region. Further, data at very high liquid fluxes is limited to only a few data sets. These limitations must be remedied before a complete understanding of flows in large pipes can be realized.

2.6.4 Boiling Two-Phase Flow in Large Diameter Pipe

For a flow with generation of void at the wall due to nucleate boiling, the drift-flux models developed by Hibiki and Ishii [14] and Kataoka and Ishii [24] have not been validated by experimental data. For such a flow condition, the distribution parameter should have a near-zero value at the beginning of the boiling two-phase-flow region. With the increase in the cross-sectional mean void fraction, the peak of the local void fraction moves from the near-wall region to the central region. This will lead to the increase in the value of the distribution parameter as the void profile develops.

In view of these basic characteristics described above, Ishii [20] proposed the following simple correlation for the distribution parameter of boiling flow.

$$C_0 = \left\{ C_\infty - (C_\infty - 1)\sqrt{\frac{\rho_g}{\rho_f}} \right\} \{1 - \exp(k\langle\alpha\rangle)\} \tag{2.16}$$

where k is a coefficient. Ishii determined C_∞ (the asymptotic value of C_0) to be 1.2 and the coefficient, k, to be -18 with experimental data taken in relatively small channels (1977). For a wall nucleate boiling in a large diameter pipe, the above model should also be sound. Equation (2.16) suggests that the void fraction would be the dominant factor for determining the distribution parameter in boiling flow. The impacts of different fluids and different pressures are included through the physical properties. The changes in the distribution parameter with void fraction and pressure are shown in Fig. 2.41 for water and steam boiling flow at saturation states, with the operating pressures of BWR and PWR highlighted.

As mentioned above, as the cross-sectional mean void fraction increases, the peak of the local radial void fraction profile moves from the near-wall region to the central region. When the void fraction is high enough, the boiling two-phase flow would have flow characteristics similar to non-boiling two-phase flow. The comparisons in Sect. 2.6.3.2 showed that the high void fraction steam-water data without boiling were well predicted by the drift-flux correlation of Kataoka and

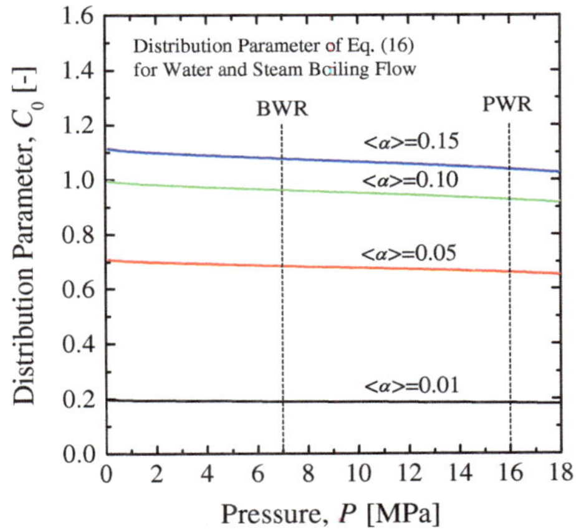

Fig. 2.41 Boiling flow distribution parameter changes with void fraction and pressure at saturation states

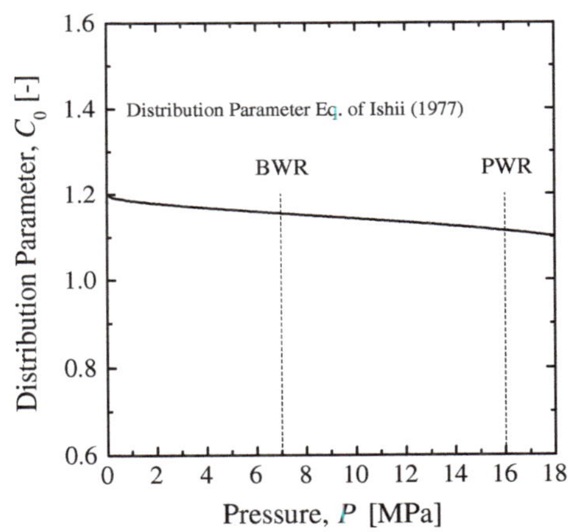

Fig. 2.42 Distribution parameter of Ishii [20] changes with pressure at saturation states

Ishii [24]. The model of Kataoka and Ishii consists of the original distribution parameter correlation of Ishii [20] and the drift velocity correlation of Kataoka and Ishii [24]. So the two correlations would be expected to be able to predict boiling two-phase flows with high void fraction. The distribution parameter and drift velocity correlations do not depend on the void fraction but change with the different fluids and different pressures through the physical properties. The changes in the distribution parameter and drift velocity with pressure are shown in Figs. 2.42 and 2.43 respectively for water and steam boiling flow at saturation states, with the operating pressures of BWR and PWR highlighted.

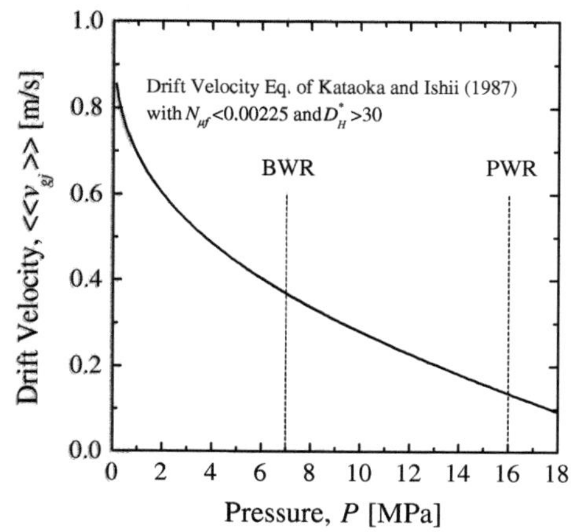

Fig. 2.43 Drift velocity of Kataoka and Ishii [24] changes with pressure at saturation states

2.7 Summary

Two-phase flows are essential in a wide variety of industrial applications, and in many of those applications large diameter channels form an integral component. To predict the performance of these systems, the ability to predict the void fraction in such large diameter channels is essential. This task is complicated, however, by the many differences between small and large diameter channels. The instability of cap bubbles leading to the lack of typical gas slugs in larger pipes leads to changes in the turbulence characteristics, flow pattern and bubble behavior, and relative velocity between the phases as the pipe diameter increases. The overall effect of these changes is that models for small diameter channels, which have been well-developed, cannot be used to predict flows in large diameter channels. To predict two-phase flows, the drift-flux model is commonly used due to its simplicity and ease of use. Many correlations have been developed using the framework provided by the drift-flux model in order to predict the behavior of two-phase flows in large diameter channels. Many of these correlations are empirically based on only a few data sets and so have limited applicability and accuracy.

In this chapter, the latest progress in two-phase flows in large diameter pipes is discussed. The flow characteristics in large diameter pipes are analyzed and summarized by using experimental data. The flow regimes and their transition criteria are discussed. The existing drift-flux models are collected and their strengths and weaknesses are noted. The up-to-date void fraction data that can be used to evaluate these models are presented. In view of the different flow regimes in the large diameter pipe, the existing drift-flux models are evaluated systematically in both low and high void fraction flows. Based on the existing database, the drift flux models of Hibiki and Ishii [14] and Kataoka and Ishii [24] seem to perform best for bubbly, cap-turbulent and churn-turbulent flow. They are recommended for use in predicting the void fraction in bubbly flow [14] and in cap-turbulent and churn-turbulent flows [24].

References

1. R.V. Baily, P.C. Zmola, F.M. Taylor, R.J. Planchet, Transport of gases through liquid-gas mixture. Paper presented at the AIChE New Orleans meeting (1956)
2. S.G. Bankoff, A variable density single-fluid model for two-phase flow with particular reference to steam-water flow. J. Heat Transf. **82**, 265–272 (1962)
3. F. Carrier, Steam separation technology under the Euroatom program. Allis-Chalmers Atomic Energy Division, report ACNP-63021 (1963)
4. B. Chexal, G. Lellouche, J. Horowitz, J. Healzer, S. Oh, The Chexal-Lellouche void fraction correlation for generalized applications. Nuclear Safety Analysis Centre of the Electric Power Research Institute, Report NSAC-139 (1991)
5. N.N. Clark, R.L. Flemmer, Predicting the holdup in two-phase bubble upflow and downflow using the zuber and findlay drift-flux model. AIChE J. **31**, 500–503 (1985)
6. N.N. Clark, R.L. Flemmer, The effect of varying gas voidage distributions on average holdup in vertical bubble flow. Int. J. Multiph. Flow **12**, 299–302 (1986)
7. J.P. Cunningham, H.C. Yeh, Experiments and void correlation for PWR small-break LOCA conditions. Trans. ANS **17**, 369–370 (1973)
8. A.I. Filiminov, M.M. Przhizhalovski, E.P. Dik, J. Petrova, The driving head in pipes with a free interface in the pressure range from 17 to 180 atm. Teploenergetika **4**, 22 (1957)
9. R.I.L. Guthrie, A.V. Bradshaw, The stability of gas envelopes trailed behind large spherical cap bubbles rising through viscous liquids. Chem. Eng. Sci. **24**, 913–917 (1969)
10. W.H. Hall, W.P. Prueter, T.L. Thorne, J.R. Wall, High-pressure steam/water void fraction profiles in a large diameter, vertical pipe with non-developed entrance flow, in *Proceedings of Thermal Hydraulics of Nuclear Steam Generators/Heat Exchangers*, Nov 27-Dec2, Chicago, IL, USA (1988)
11. A. Hashemi, J.H. Kim, J.P. Sursock, Effect of diameter and geometry on two-phase flow regimes and carry-over, in *Proceedings of the 8th International Heat Transfer Conference*, San Francisco, USA, pp. 2443–2451 (1986)
12. T. Hibiki, M. Ishii, Experimental study on hot-leg U-bend two-phase natural circulation in a loop with a large diameter pipe. Nucl. Engng. Des. **195**, 69–84 (2000)
13. T. Hibiki, M. Ishii, Effect of inlet geometry on hot-leg U-bend two-phase natural circulation in a loop with a large diameter pipe. Nucl. Engng. Des. **203**, 209–228 (2001)
14. T. Hibiki, M. Ishii, One-dimensional drift-flux model for two-phase flow in a large diameter pipe. Int. J. Heat Mass Transf. **46**, 1773–1790 (2003)
15. T. Hibiki, M. Ishii, Z. Xiao, Axial interfacial area transport of vertical bubbly flows. Int. J. Heat Mass Transf. **44**, 1869–1888 (2001)
16. J.H. Hills, The operation of a bubble column at high throughputs I. Gas holdup measurements. Chem. Eng. J. **12**, 89–99 (1976)
17. Y. Hirao, K. Kawanishi, A. Tsuge, T. Kohriyama, Experimental study on drift-flux correlation formulas for two-phase flow in large diameter tubes, in *Proceedings of the 2nd International Topical Meeting on Nuclear Power Plant Thermal-Hydraulics and Operations,* Tokyo, Japan, pp. 1–88–1–94 (1986)
18. J.T. Hsu, M. Ishii, T. Hibki, Experimental study on two-phase natural circulation and flow termination in a loop. Nucl. Eng. Des. **186**, 395–409 (1998)
19. Y. Inoue, Measurement of interfacial area concentration of gas-liquid two-phase flow in a large diameter pipe. MS thesis, Kyoto University, 2001
20. M. Ishii, One-dimensional drift-flux model and constitutive equations for relative motion between phases in various two-phase flow regimes. ANL-77-47, USA (1977)
21. M. Ishii, T.C. Chawla, Local drag Laws in dispersed two-phase flow, Argonne National Lab. Report, ANL-79-105 (1979)
22. M. Ishii, T. Hibiki, *Thermo-Fluid Dynamics of Two-Phase Flow*, 2nd edn. (Springer, New York, 2010)

23. M. Ishii, G. Kocamustafaogullari, Maximum fluid particle size for bubbles and drops, in *Proceedings of ASME Winter Annual Meeting*, Miami Beach, FL, USA, FED-vol. 29, pp. 99–107 (1985)
24. I. Kataoka, M. Ishii, Drift-flux model for large diameter pipe and new correlation for pool void fraction. Int. J. Heat Mass Transf. **30**, 1927–1939 (1987)
25. I. Kataoka, M. Ishii, A. Serizawa, Local formulation and measurements of interfacial area concentration in two-phase flow. Int. J. Multiph. Flow **12**, 505–529 (1986)
26. T.J. Liu, S.G. Bankoff, Structure of air-water bubbly flow in a vertical pipe—I. Liquid mean velocity and turbulence measurements. Int. J. Heat Mass Transf. **36**, 1049–1060 (1993a)
27. T.J. Liu, S.G. Bankoff, Structure of air-water bubbly flow in a vertical pipe—II. Void fraction, bubble velocity and bubble size distribution. Int. J. Heat Mass Transf. **36**, 1061–1072 (1993b)
28. K. Mishima, M. Ishii, Flow regime transition criteria for upward two-phase flow in vertical tubes. Int. J. Heat Mass Transf. **27**, 723–737 (1984)
29. M. Murase, H. Suzuki, T. Matsumoto, M. Naitoh, Countercurrent gas-liquid flow in boiling channels. J. Nucl. Sci. Technol. **23**, 487–502 (1986)
30. A. Ohnuki, H. Akimoto, An experimental study on developing air-water two-phase flow along a large vertical pipe: effect of air injection method. Int. J. Multiph. Flow **22**, 1143–1154 (1996)
31. A. Ohnuki, H. Akimoto, Experimental study on transition of flow pattern and phase distribution in upward air–water two-phase flow along a large vertical pipe. Int. J. Multiph. Flow **26**, 367–386 (2000)
32. A. Ohnuki, H. Akimoto, Model development for bubble turbulent diffusion and bubble diameter in large vertical pipes. J. Nucl. Sci. Technol. **38**, 1074–1080 (2001)
33. A. Ohnuki, H. Akimoto, Y. Sudo, Flow pattern and its transition in gas-liquid two-phase flow along large vertical pipe, in *Proceeding of The 2nd International Conference on Multiphase Flow '95-Kyoto*, vol. 3, pp. FT1_17–23 (1995)
34. N.K. Omebere-Iyari, B.J. Azzopardi, D. Lucas, M. Beyer, H.M. Prasser, The characteristics of gas/liquid flow in large risers at high pressure. Int. J. Multiph. Flow **34**, 461–476 (2008)
35. V.I. Petukhov, V.A. Kolokoltsen, Effect of liquid viscosity on droplet entrainment and volumetric air content. Teploenergetica **12**, 30 (1965)
36. P. Sawant, J.P. Schlegel, S. Paranjape, B. Ozar, T. Hibiki, M. Ishii, Flow regime identification in large diameter pipe, in *Proceedings of the 16th International Conference on Nuclear Engineering*, Orlando, Florida, USA, pp. 341–351 (2008)
37. J.P. Schlegel, T. Hibiki, M. Ishii, Development of a comprehensive set of drift-flux constitutive models for pipes of various hydraulic diameters. Prog. Nucl. Energy **52**, 666–677 (2010)
38. J.P. Schlegel, P. Sawant, S. Paranjape, B. Ozar, T. Hibiki, M. Ishii, Void fraction and flow regime in adiabatic upward two-phase flow in large diameter vertical pipes. Nucl. Eng. Des. **239**, 2864–2874 (2009)
39. J.P. Schlegel, S. Miwa, S. Chen, T. Hibiki, M. Ishii, Experimental study of two-phase flow structure in large diameter pipes. Exp. Therm. Fluid Sci. **41**, 12–22 (2012)
40. A. Serizawa, I. Kataoka, I. Michiyoshi, Turbulence structure of air-water bubbly flow-I. Measuring techniques. Int. J. Multiph. Flow **2**, 221–233 (1975)
41. A. Serizawa, I. Kataoka, Turbulence suppression in bubbly two-phase flow. Nucl. Eng. Des. **122**, 1–16 (1990)
42. M.E. Shawkat, C.Y. Ching, M. Shoukri, Bubble and liquid turbulence characteristics of bubbly flow in a large diameter vertical pipe. Int. J. Multiph. Flow **34**, 767–785 (2008)
43. X. Shen, R. Matsui, K. Mishima, H. Nakamura, Distribution parameter and drift velocity for two-phase flow in a large diameter pipe. Nucl. Eng. Des. **240**, 3991–4000 (2010a)
44. X. Shen, K. Mishima, H. Nakamura, Two-phase phase distribution in a vertical large diameter pipe. Int. J. Heat Mass Transf. **48**, 211–225 (2005)

45. X. Shen, K. Mishima, H. Nakamura, Flow-induced void fraction transition phenomenon in two-phase flow, in *Proceedings of 8th International Conference on Nuclear Engineer (ICONE18)*, Xian, China, Paper No. ICONE18-29630 (2010b)
46. X. Shen, Y. Saito, K. Mishima, H. Nakamura, Methodological improvement of an intrusive four-sensor probe for the multi-dimensional two-phase flow measurement. Int. J. Multiph. Flow **31**, 593–617 (2005)
47. X. Shen, Y. Saito, K. Mishima, H. Nakamura, A study on the characteristics of upward air-water two-phase flow in a large pipe. Exp. Therm. Fluid Sci. **31**, 21–36 (2006)
48. D.G. Shipley, Two phase flow in large diameter pipes. Chem. Eng. Sci. **39**, 163–165 (1984)
49. T.R. Smith, Two-phase flow regime identification in large diameter pipes. MS thesis, Purdue University, 1999
50. T.R. Smith, Two-group interfacial area transport equation in large diameter pipes. Dissertation, Purdue University, 2002
51. T.R. Smith, J.P. Schlegel, T. Hibiki, M. Ishii, Two-phase flow structure in large diameter pipes. Int. J. Heat Fluid Flow **33**, 156–167 (2012)
52. M.A. Styrikovich, A.V. Surnov, J.G. Vinokour, Experimental data on hydrodynamics of two-phase mixture. Teploenergetica **8**, 56 (1961)
53. M.A. Styrikovich, S.S. Kutateladze, Hydrodynamics of liquid-gas system. Energy 196 (1976)
54. X. Sun, T. Smith, S. Kim, M. Ishii, J. Uhle, Interfacial area of bubbly flow in a relatively large diameter pipe. Exp. Therm. Fluid Sci. **27**, 97–109 (2002)
55. Y. Taitel, D. Bornea, A.E. Dukler, Modeling flow pattern transitions for steady upward gas-liquid flow in vertical tubes. AIChE J. **26**, 345–354 (1980)
56. J.F. Wilson, R.J. Grenda, J.F. Patterson, Steam volume fraction in a bubbling two-phase mixture. Trans. ANS **4**, 356–357 (1961)
57. K. Yoneda, A. Yasuo, T. Okawa, Flow structure and bubble characteristics of steam-water two-phase flow in a large diameter pipe. Nucl. Engng. Des. **217**, 267–281 (2002)
58. N. Zuber, J.A. Findlay, Average volumetric concentration in two-phase flow systems. J. Heat Transf. **87**, 453–468 (1965)
59. X. Shen, T. Hibiki, H. Nakamura, Developing structure of two-phase flow in a large diameter pipe at low liquid flow rate. Int. J. Heat Fluid Flow **34**, 70–84 (2012)
60. H. Prasser, Evolution of interfacial area concentration in a vertical air–water flow measured by wire–mesh sensors. Nucl. Eng. Des. **237**, 1608–1617 (2007)
61. X. Shen, K. Mishima, H. Nakamura, Error reduction, evaluation and correction for the intrusive optical four-sensor probe measurement in multi-dimensional two-phase flow. Int. J. Heat Mass Transf. **51**, 882–895 (2008)
62. X. Shen, K. Mishima, H. Nakamura, A method for measuring local instantaneous interfacial velocity vector in multi-dimensional two-phase flow. Int. J. Multiph. Flow **34**, 502–509 (2008)

Chapter 3
Physical Water Treatment Using Oscillating Electric Fields to Mitigate Scaling in Heat Exchangers

Leonard D. Tijing, Cheol Sang Kim, Dong Hwan Lee and Young I. Cho

Abstract This chapter presents an environment-friendly method to mitigate scaling in heat exchangers. A new physical water treatment (PWT) using high-frequency oscillating electric fields produced directly in water was used to mitigate scaling of heat transfer surfaces. The new method of using high-frequency oscillating electric fields directly in water is a major improvement over the previous PWT methods (i.e., low electric field strength, about ~ 1 mV/cm, and low allowable frequency, ~ 2 kHz). Both artificial and natural hard water at varying calcium carbonate hardness were used. Different combinations of voltages and frequencies were investigated to get the optimum values for the mitigation of scaling. It is hypothesized that the oscillating electric fields in the present PWT method precipitate the dissolved mineral ions such as calcium to mineral salts in bulk water. As the mineral ions continue to precipitate and adhere on the surfaces of the suspended particles, the particles grow in size and adhere to the solid heat transfer surface in the form of soft sludge or particulate fouling. This type of fouling is believed to be easily removed by shear forces created by flow than those deposits produced from

L. D. Tijing (✉) · C. S. Kim · D. H. Lee
Division of Mechanical Design Engineering, Chonbuk National University, Jeonju, Jeonbuk 561-756, Republic of Korea
e-mail: ltijing@gmail.com; ltijing@jbnu.ac.kr

L. D. Tijing
School of Civil and Environmental Engineering, University of Technology, Sydney (UTS), P.O. Box 123, Broadway, NSW 2007, Australia

C. S. Kim
e-mail: chskim@jbnu.ac.kr

D. H. Lee
e-mail: 0311dhlee@jbnu.ac.kr

Y. I. Cho
Department of Mechanical Engineering and Mechanics, Drexel University, Philadelphia, PA 19104, USA
e-mail: choyi@drexel.edu

L. Cheng (ed.), *Frontiers and Progress in Multiphase Flow I*,
Frontiers and Progress in Multiphase Flow, DOI: 10.1007/978-3-319-04358-6_3,

the precipitation of mineral ions directly on the solid heat transfer surface, i.e., precipitation fouling. The new PWT method using oscillating electric fields presents a valid tool to mitigate scaling in heat exchangers from cooling water. The work in this book is based from the PhD dissertation of the first author at the Division of Mechanical Design Engineering at Chonbuk National University. Section 3.1 presents an overview of mineral fouling and the different methods to mitigate the fouling formation in heat exchangers, focusing on physical water treatment. Sections 3.2, 3.3, 3.4 and 3.5 give in detail the experimental work and discussion of the use of oscillating electric fields as a means to mitigate mineral fouling in a double-pipe heat exchanger. Section 3.6 summarizes the present study.

Nomenclature

A_o	Outer surface area of copper tube (m^2)
c_p	Specific heat of water (J/kg K)
d	Diameter of copper tube (m)
D	Diameter of quartz crystal (m)
f	Frequency (Hz)
H	Height (m)
L	Length (m)
$\dot{m}$	Mass flow rate (kg/s)
Mw	Molecular weight (g/mol)
Q	Heat transfer rate (W)
R_f	Fouling resistance (m^2K/W)
ΔT	Temperature difference (°C)
U	Overall heat transfer coefficient (W/m^2 K)
W	Width (m)

Greek symbol

θ Diffraction angle (°)

Subscripts

c	Cold side
f	Fouled state
h	Hot side
ini	Initial clean state
in	Inner
lmtd	Log-mean-temperature difference (°C)
out	Outer
q	Quartz crystal
t	Tube (copper)

3.1 Introduction

3.1.1 Mineral Fouling

Scale formation on heat transfer surfaces of heat exchangers presents a critical concern in industrial operation and maintenance [1, 2]. The scale formation is commonly called as mineral fouling, which involves precipitation and particulate fouling mechanisms. Circulating water in a cooling tower contains excessive mineral ions especially calcium and magnesium ions as the evaporation of pure water is the primary mode of heat rejection, leaving those mineral ions behind [3, 4]. These dissolved minerals when supersaturated usually precipitate out of the solution when exposed to high temperature because of their inverse solubility property and changes in pH of water [1–22]. In particular, calcium ions reacting with carbonate ions in water form layers of calcium carbonate ($CaCO_3$) deposit on heat transfer surfaces, decreasing the efficiency of heat exchangers because of the insulating effect of the deposits [23, 24]. Scale deposits have very low thermal conductivity. For example, calcium carbonate has a thermal conductivity of only 2.9 W/mK [13], while that of the metal surface made of copper has a conductivity of 401 W/mK [25]. Furthermore, the formed deposits reduce the flow area, thus requiring more pumping power to achieve the flow rate of water corresponding to the clean state [1–9, 21]. A 0.8-mm layer of $CaCO_3$ scale can increase the energy use by about 10 % [10].

The scaling propensity depends on several factors such as the kind of salt used, concentration, pH, and the operating conditions [26]. Precipitation can either be produced in bulk water or at the surface depending on which area reflects the salt saturation temperature. The scales produced from mineral fouling are often hard and tenacious. There are very high chances for mineral fouling to occur in industrial equipment especially those that utilize "natural" waters for cooling reasons [27]. Steinhagen et al. [11] showed in a New Zealand survey that 90 % of heat exchangers had fouling problems. The cost for cleaning and maintenance of heat exchangers due to fouling in industries reaches billions of dollars each year [27]. If one can lessen or completely eradicate fouling on heat transfer surfaces, this not only increases heat exchanger efficiency, but also reduces the expenses associated with cleaning of fouled heat exchangers. Furthermore, as the fouling can be mitigated, the cycles of concentration (COC) can be increased, resulting in water savings by reduced make-up and blowdown [9, 12–14].

Calcium carbonate is one of the most common scales found in cooling-water applications [28–32]. It exists in three crystal phases but the two most common are aragonite and calcite. Aragonite has a specific gravity of 2.95 while calcite has 2.71. The aragonite form is denser than the calcite form and is considered to be the more troublesome type between the two. It is formed at above 30 °C while calcite is generally formed at room temperature. Calcite embodies a round structure and hexagonal shape making it not too adherent to the heat transfer surface. On the other hand, aragonite has sharp edges similar to needle shape so that it tends to

Table 3.1 Comparison of the two common crystal structures of $CaCO_3$: calcite and aragonite

	Calcite		Aragonite	
Specific gravity	2.71		2.94	
Adhesiveness	Less adherent		Adherent	
Crystal form	Hexagonal		Orthorhombic	
Microscopic visualization	Round shape		Needle shape	
Formation temperature	Room temperature (below 30 °C)		Above 30 °C	

adhere more on the surface, and quite difficult to remove [33]. Table 3.1 shows the differences between the two types [5, 6].

The formation of $CaCO_3$ scale is a multistage process. When rainwater and groundwater come in contact with limestone or chalk, they absorb mineral ions (mostly calcium), which contributes to the hardness of natural water. When this natural hard water enters pipes, they can form scaling. The typical calcium concentration in natural water is less than 15 mg/l, but when water is exposed to carbonate rocks, the concentration could reach 30–100 mg/l [34, 35]. The scaling is formed when these dissolve calcium ions are exposed to changes in temperature, where at high temperature, the dissolved ions are forced out of the solution and form into scales [36].

The chemical reaction for the precipitation of $CaCO_3$ generally includes three chemical reactions among ions within the water [36]. As the temperature in the heat exchanger gets higher, the water ion contents especially calcium and bicarbonate ions dissociate from the water molecules surrounding them. This happens at a supersaturated condition. The first step involves the dissociation of bicarbonate ions into hydroxide ions and carbon dioxide in liquefied form (Eq. 3.1.1). The production of bicarbonate ions consequently increases the water pH. When the aqueous carbon dioxide evaporates into the open air as time goes by, it further increases the pH of water at above 7.0. At this instance, the excess amount of hydroxide ions reacts further with the other bicarbonate ions thereby producing carbonate ions (Eq. 3.1.2). The heat transfer surface at this time has become negatively charged due to the transport of the excess hydroxide ions, thereby attracting the positively charged calcium ions and eventually making an "electric double layer" near the surface [6]. Upon the completion of the second reaction, a spontaneous reaction occurs between calcium and carbonate ions, producing the problem-bound scales as calcium carbonate (Eq. 3.1.3). The following presents the reactions discussed above [6, 36]:

$$HCO_3^-{}_{(aq)} \leftrightarrow OH^-{}_{(aq)} + CO_{2(aq)} \tag{3.1.1}$$

$$OH^-{}_{(aq)} + HCO_3^-{}_{(aq)} \leftrightarrow CO_3^{2-}{}_{(aq)} + H_2O \tag{3.1.2}$$

$$Ca^{2+}{}_{(aq)} + CO_3^{2-}{}_{(aq)} \leftrightarrow CaCO_{3(s)} \tag{3.1.3}$$

3.1.2 Fouling Prevention and Mitigation Methods

There are several methods to help lessen, control, or prevent the scaling of heat exchangers, which can be primarily divided into two: chemical and non-chemical methods. In the past, the use of scale-inhibiting chemicals like chlorine, brominated compounds, organic and inorganic phosphates as well as polymer scale inhibitors was the best choice for the control of mineral fouling. Although it had a high success rate, there were also many disadvantages and concerns in their use. Aside from the high cost of chemicals, more stringent environmental laws increased the costs associated with their storage, handling and disposal [37, 38]. These chemicals or accumulated chemical residues over a long period of time also pose concern to the environment with accidental spills. Thus, there is a need for a new approach that is safe and clean from both environmental and cost points of view in the maintenance of heat exchangers [5, 7, 10, 13, 39, 40].

Physical water treatment (PWT) is a non-chemical method to mitigate mineral fouling with the use of electric or magnetic fields, catalytic surfaces, ultrasounds, or sudden pressure changes. Numerous studies have been reported for the effectiveness of the use of permanent magnets [5, 6, 10, 13, 19, 41–45], solenoid-coils [5, 6, 46–51], catalytic materials like copper, zinc [52–55], tourmaline [56], and titanium [18], and ultrasounds [12]. Al-Qahtani [57] reported that the magnetic fields increased the formation of crystals at the bulk fluid rather than at the surface and at the same time, accelerated the coagulation-flocculation processes of suspended particles in water [6, 58–61]. Sohnel and Mullin [62] concluded that magnetic or electric fields directly influenced bulk precipitation and also by hydrodynamic cavitation. Parsons et al. [41] concluded their research stating that the mechanism of PWT may be due to microcontaminant effects or by interfacial phenomena. Lisitsin et al. [63] reported a decreased bulk precipitation of calcium carbonate and reduced scaling of reverse osmosis membranes when a 2-ppm concentration of zinc was dosed in moderately hard water system. Many studies with the use of magnetic and electric fields suggest the formation of calcium carbonate particles in the bulk of the scaling water rather than directly on the walls of the distribution pipes and heated surfaces, thus diminishing the chances of strong adherence of scales on the pipe walls [5, 6, 14, 15, 50, 64, 65]. Alimi et al. [65] found that magnetic treatment of hard water using permanent magnets increases the amount of precipitate in the bulk water and favors homogeneous nucleation. They concluded that the solution pH, flow rate and the residence time of flowing water are important parameters for bulk precipitation. Kim et al. [66] showed the effect of electric fields prior to microfiltration on membrane fouling, killing of bacteria and particle coagulation. Their results showed more particle coagulation, reduced membrane fouling, and increased number of inactivated bacteria when the wastewater was treated with electric fields. The electric field increases the chances of ions to collide with each other, thus increasing their reactivity, and precipitate in the bulk solution faster and more effective [46]. Some of the parameters studied on the effect of PWT to scaling kinetics of calcium carbonate were changes in water chemistry, hydrodynamic conditions, surface material and surface finish [54, 67]. Cho et al. conducted a thorough study on the exact mechanism of PWT. Among the reviewed mechanisms are the direct effect of electric and magnetic fields on the surface potential energy and chemical reaction rate, separation of charges, and heterogeneous catalysis at the surface. They concluded that the main mechanism of PWT is due to bulk precipitation via heterogeneous catalysis. They reported that particulates are formed due to bulk precipitation which subsequently went to the heat transfer surface as a soft sludge [60]. They further related that PWT will not work unless there is enough cleaning force to dislodge the particulate deposition [59, 67]. The use of filtration to trap the precipitated particles is suggested in order to maximize the mitigating effect of PWT. Their observations were based on using solenoid coils and permanent magnets that induce electric and magnetic fields.

In this study, high frequency oscillating electric fields generated between two graphite electrode plates directly in water were used to mitigate mineral fouling in a double-pipe heat exchanger [68]. The new method of using oscillating electric fields directly in water is a major improvement over the above-mentioned PWT methods because these previous PWT methods produce induced electric fields in water so that the field strength is often very small (~ 1 mV/cm) and the maximum frequency of the field is limited to approximately 2 kHz due to involved physics laws such as Faraday's law [5, 6, 69–80]. The fouling thermal resistance of the fouled deposit was determined and the chemistry of water was analyzed during the calcium carbonate scaling process using natural and artificial hard water. Furthermore, the scale deposits were characterized by scanning electron microscopy (SEM) and X-ray diffraction (XRD), digital photographs of the fouled tubes, particle count in the bulk water, and the concentration of calcium on calcium carbonate scale formed on a heated surface measured by inductively coupled plasma-mass spectrometry (ICP-MS, 7500a, Agilent, USA). Table 3.2 gives the different fouling researches done in the past with varying experimental conditions such as heat flux, flow velocity, water concentration and types of foulant. In the present study, we conducted fouling experiments by varying the heat fluxes, flow velocities at the cold side, and the concentration of calcium carbonate in water.

3.1.3 Physical Water Treatment Using Oscillating Electric Fields

Figure 3.1 shows the PWT device using two graphite electrode plates to produce oscillating electric fields in the present study [68]. The two graphite plates ($L = 155 \pm 0.05$ mm; $W = 7.65 \pm 0.05$ mm; and $H = 35 \pm 0.05$ mm) were placed in parallel to each other inside a cylindrical Teflon block, forming a rectangular cross-section. The distance between the two graphite electrode plates was fixed at 20 ± 0.05 mm. In order to ensure that the plate electrodes did not touch

Table 3.2 Comparison of the present and previous fouling studies

References	Heat flux (kW/m^2)	Flow velocity (m/s)	Concentration (mg/l)	Foulant
Hasson et al. [32]	1.6	0.25–0.82	110.575	Calcium
Kim and Webb [60]	13	0.8–1.82	1,500	Aluminum oxide/ferric oxide
Somerscales et al. [61]	28–52	0.9–1.0 1.4–1.5	2,500	MgO
Nasrazadani and Chao [81]	137	1–1.98	300–450	Calcium
Sheikholeslani and Watkinson [82]	120–220	0.3–0.8	603–700	Calcium
Morse and Knudsen [83]	276	1.0	490–650	TDS
Helalizadeh et al. [22]	100–400	0.5–2.0	1.0–2.5/ 0.25–1.0	Calcium sulphate/ calcium carbonate
Cho et al. [5]	380–485	1.2–1.5	600–1500	Total hardness
Lee et.al. [50]		0.3–0.8	3,600	Calcium (conductivity)
Tijing et al. [18]	125–140	0.3–0.5	6,150	Calcium (conductivity)
Shahryari and Pakshir [50]		0.5–1.3	318	Calcium
Li and Webb [3]		1.07	1,300	Aluminum oxide
Quan et al. [48]		1.2	300	Calcium carbonate
Kazi et al. [58]		0.1	3,600	Calcium sulphate
Paakkonen et al. [84]	47–63	0.2–0.4		Calcium carbonate
Present study	2.9–5.8	0.3–1.0	350–550	Calcium carbonate

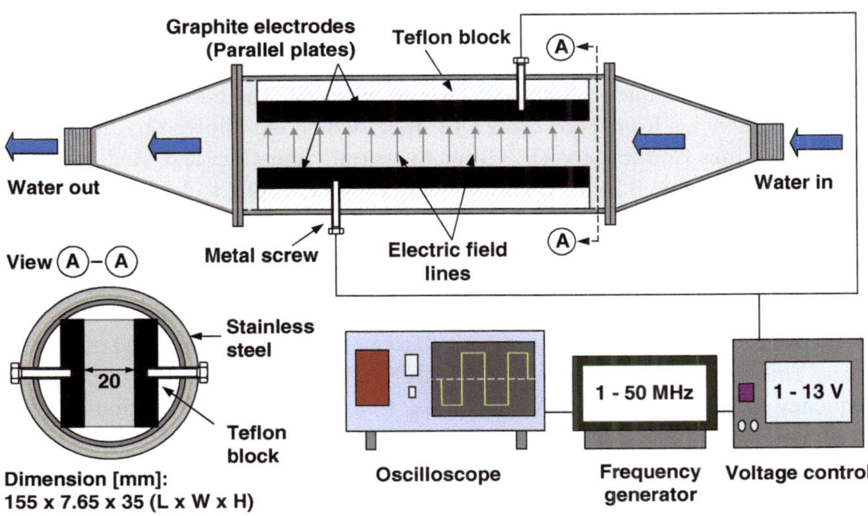

Fig. 3.1 The present PWT device using two graphite electrode plates [68]

Table 3.3 FCC ISM frequencies [85]

ISM frequency	Tolerance	Wavelength
6.78 MHz	±15.0 kHz	145 ft (44.4 m)
13.56 MHz	±7.0 kHz	72.6 ft (22.1 m)
27.12 MHz	±163.0 kHz	36.3 ft (11.1 m)
40.68 MHz	±20.0 kHz	24.2 ft (7.37 m)
915 MHz	±13.0 MHz	1.08 ft (32.8 cm)
2,450 MHz	±50.0 MHz	4.82 in (12.2 cm)
5,800 MHz	±75.0 MHz	2.03 in (5.17 cm)
24,125 MHz	±125.0 MHz	0.49 in (1.24 cm)
61.25 GHz	±250.0 MHz	0.19 in (4.99 cm)
122.50 GHz	±500.0 MHz	0.09 in (0.24 cm)
245.00 GHz	±1.0 GHz	0.05 in (0.12 cm)

Note Europe also uses 433.92 MHz ± 870 kHz

any metal parts in the case, a cylindrical Teflon block with square cross-section which was tightly fit between the casing and the electrodes was used. The present device was also specially designed so that the plate electrodes could be screwed at opposite sides from the outside of the device through screw ports, which served as the connections for the power supply, thus eliminating any direct contact of water to the electric wire for power supply. The present PWT device used a stainless steel cylinder casing where detachable diverging and converging head sections were used at the inlet and outlet of the casing for smooth transitions of flow, respectively. O-rings were used between the main cylinder casing and the detachable heads, and locked with screw-type locks. The power supply included a voltage controller (100 mA, ±15 V), a programmable waveform generator (1–50 MHz), and an oscilloscope. The waveform generator could generate square wave signals in a frequency range of 1 Hz to 50 MHz. The current study utilized various frequencies from 1 kHz to 27.12 MHz, which are allowed by the Federal Communications Commission (FCC) for Industrial, Scientific, and Medical (ISM) equipment (see Table 3.3) [85]. Test voltages ranging from 2 to 13 V were used to determine the optimum voltage. We did not control the current of the system but was maintained at a minimum depending on the voltage used. Our focus was on the effect of high-frequency electric fields on the fouling formation on heated surfaces, thus current is not a parameter.

The present PWT method utilized oscillating electric fields produced directly in water. This allows oscillating electric fields at a significantly higher strength and frequency than the previous PWT methods, which used indirect methods such as permanent magnets, solenoid coils and electrostatic device. These devices produce induced electric fields indirectly, where the production of electric field is governed by basic physics law such as Lorentz law or Faraday's law. Due to the induction process, the strength and magnitude of the induced electric field is in the order of 1 mV at or less than 3,000 Hz. Of note that the higher frequency beyond 3,000 Hz is desirable for more efficient operation of the PWT, but the aforementioned basic physics laws prevent us from using a higher frequency in the previous PWT devices. There are three ways mineral ions precipitate inside the heat exchanger: (a) directly on the heat transfer surface where the temperature is high, (b) on the surface of solid particles suspended in water [86], and (c) in bulk water [10]. It is hypothesized that the oscillating electric fields in the present PWT method precipitate the dissolved mineral ions such as calcium to mineral salts in bulk water. As the treated water enters the heat exchanger, the temperature of water increases, and the solubility of the calcium ions decreases and calcium ions precipitate and adhere on the solid surface of heat exchanger. As the mineral ions continue to precipitate and adhere on the surfaces of the suspended particles, the particles grow in size and adhere to the solid heat transfer surface in the form of soft sludge or particulate fouling. This type of fouling is believed to easily be removed by shear forces created by flow than those deposits produced from the precipitation of mineral ions directly on the solid heat transfer surface, i.e., precipitation fouling [10]. The present PWT method utilized a low voltage not only to provide safety in equipment and to personnel, but also to minimize the electric power consumption for energy efficiency.

3.1.4 Objectives

Calcium carbonate fouling is a rampant occurrence for cooling-tower applications. Thus, the present study simulated $CaCO_3$ fouling by preparing natural hard water through cooling-tower recirculation process and artificial hard water through mixing of an appropriate amount of sodium bicarbonate ($NaHCO_3$) and calcium chloride ($CaCl_2$) powders into distilled water to obtain the desired water hardness. The present study used oscillating electric fields between two graphite electrode plates positioned parallel to each other at different voltages (2–13 V) and at different frequencies (1 kHz–27.12 MHz). The main objective of the present study was to investigate the feasibility of oscillating electric fields in mitigating $CaCO_3$ fouling in heat exchangers. The specific aims of the present study were to determine the values of voltage and frequency of the oscillating electric fields that would provide the best results for $CaCO_3$ fouling mitigation, and to investigate the effects of water hardness and flow velocity on the efficacy of the oscillating electric fields in mitigating $CaCO_3$ fouling at the optimum voltage and frequency.

3.2 Bulk Precipitation Mechanism of PWT

Many studies hypothesized that colloidal particles are formed in the bulk solution due to PWT [6, 67, 70, 74]. Parker and Lipus et al. [74] reported that magnetic water treatment produces nucleation sites in the bulk water thus precipitating in the bulk rather than on the heat transfer surface. Gehr et al. [87] concluded that

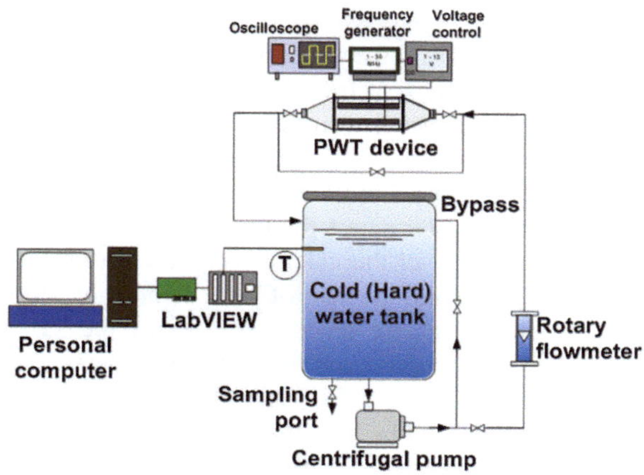

Fig. 3.2 Schematic layout of the bulk precipitation test experimental set-up

magnetic treatment increases the coagulating tendency of suspension in water. This section presents an investigation to the bulk precipitation hypothesis of PWT by measuring the number of suspended particles over time in the circulating water samples, which were concentrated natural hard water. The present study tested the present PWT at room temperature and at two different flow rates: 0.500 and 0.800 m^3/h (500 and 800 l/h, respectively).

Figure 3.2 shows an experimental set-up used in the present study. It was composed of a hard water reservoir, a centrifugal pump, a flow meter, PWT device, thermocouples, valves, and a data acquisition system (DAS). Concentrated hard water was placed in the hard water reservoir and continuously re-circulated in the system with and without PWT device. Two thermocouples were positioned inside the reservoir to monitor the temperature of the hard water in the reservoir. The thermocouples were connected to a real-time DAS using LabVIEW (National Instruments) program.

3.2.1 Preparation of Natural Hard Water Using a Cooling-Tower

Figure 3.3 shows a schematic diagram of a closed-loop cooling tower where natural hard water from a well was re-circulated to produce concentrated hard water as utilized in the study. As pure water evaporated in the cooling tower, the mineral ions such as calcium and magnesium were left behind, which were collected in the reservoir. Thus, with the recirculation of natural hard water in the mini-cooling tower, more and more mineral ions accumulated in the reservoir such that its concentration was gradually increasing until the desired electrical conductivity value of 2,300–2,700 μS/cm was produced. The properties of the present natural hard water were: pH = 8.0–8.2; hardness = 140–160 mg/l; total alkalinity = 170–180 mg/l;

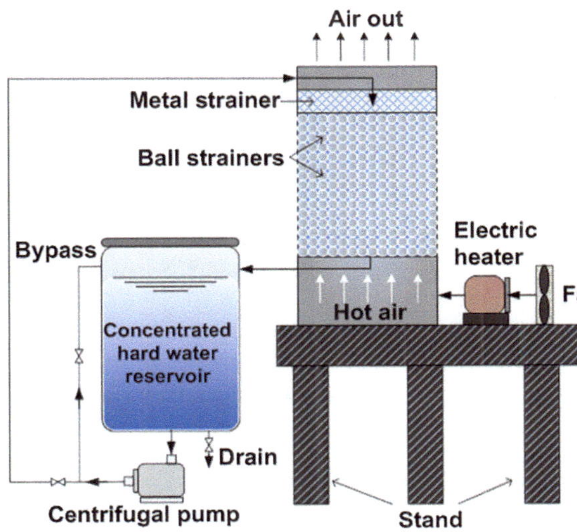

Fig. 3.3 Sketch of the present cooling tower recirculation system

silica = 15–17 mg/l; and sulfate = 110–112 mg/l. It took several days to obtain the desired concentration. A total of 0.700 m^3 (700 l) of concentrated hard water was prepared and kept in reservoir and used for the whole duration of the experiment. Each test (i.e., no-treatment case and PWT-treatment case) had 0.030 m^3 (30 l) of concentrated hard water in each reservoir. The recirculation system was consisted of a blower supplying air to the cooling tower, a heater to heat the air that was supplied, a pump to circulate the water, and the cooling tower itself filled with Styrofoam balls used as strainers.

3.2.2 Particle Number and Size Measurement

The number of particles was measured as a function of the particle size using a light-blockage technique (i.e., shadowgraphy) in a particle counter (Universal Size Distribution Analysis-System, PMT-2120-PAMAS). With the use of this light-blockage technique, it guarantees the detection of every single particle according to size and it counts on this precise principle and high resolution. Prior to each count, the particle counter machine was calibrated using distilled water as its base. The water samples collected from the present experiment were continuously stirred by an automatic stirrer in the particle counter while reading was being conducted. The schematic of particle counter is shown in Fig. 3.4. The water samples at 10 ml were placed between a light source and a detector, and the shadows cast produced by particles were measured by a photodiode. The particle sizes measured were in the range of 1–280 μm which were automatically detected by the particle counter.

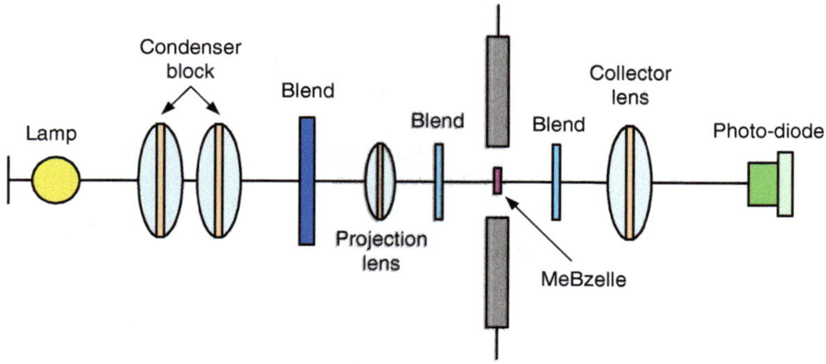

Fig. 3.4 Schematic of the particle count machine used in the study

3.2.3 Experimental Procedure

Two tests were simultaneously carried out in the present study: no-treatment (without PWT device) case served as the reference case, and PWT-treatment case (with PWT device). Tests were started using a flow rate of 500 l/h, and the other tests were conducted using a flow rate of 800 l/h. Before starting each experiment, the hard water reservoirs were thoroughly cleaned in order to minimize the presence of small particles from the previous test that could affect the particle count test. The concentrated hard water was filtrated for 12–16 h using a 50-μm cartridge filter prior to each test. Concentrated hard water of 30 l was then filled in the reservoirs (for no-treatment and PWT-treatment cases). The tests were carried out at the room temperature (20–24 °C). The temperature of the solution was continuously measured by a type-T thermocouple (Sentech, Korea) connected to a LabVIEW program for real-time data acquisition. Water samples for each case were collected at four different times during the test. The initial sample was collected at the start of the experiment. The next three water samples were collected at 18, 48, and 66 h of circulation.

3.2.4 Results for Number of Particles Versus Particle Size

Figure 3.5 shows the number of particles per volume of 10 ml as a function of particle size for the no-treatment and PWT-treatment cases at flow rates of 500 and 800 l/h. Recall that the water used in the present study was concentrated natural hard water made from continuous recirculation in a mini-cooling tower. The water properties did not change much during the whole experiment. The results given in both Fig. 3.5a and b depict in general that the number of particles less than 8 μm were significantly greater than those of larger particles (greater than 8 μm). For example, at a flow rate of 500 l/h for no-treatment case (see Fig. 3.5a,

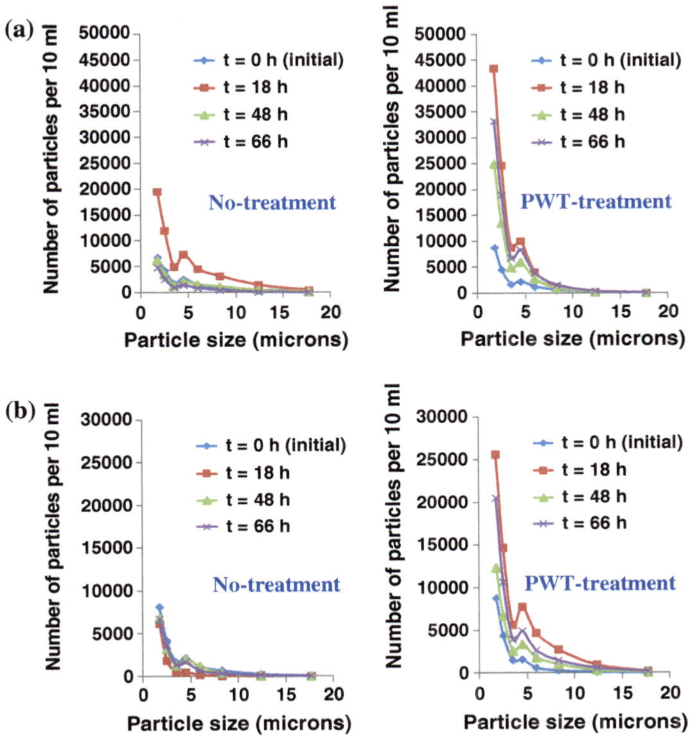

Fig. 3.5 Number of particles versus particle size at **a** 500 and **b** 800 l/h for no-treatment and PWT-treatment cases

no-treatment), the number of 1.5-μm size particles was 8,145 at day 0 as compared to 483 for particles greater than 8 μm. Both Fig. 3.5a and b obviously show a big increase in the number of particles below 8-μm size for the PWT-treatment cases as compared to the no-treatment cases for both flow rates (i.e., 500 and 800 l/h). For the PWT-treatment case, both results showed similar trends in the number of particles suspended in water. After 18 h of operation, the number of particles significantly increased which can be attributed to the effect of the PWT-treatment. As the operation continued, the precipitated particles collided with each other, producing bigger particles thereby reducing the total number of particles. At 66 h, some of the bigger particles might have reached a point where they settle at the bottom of the reservoir or adhere to pipe surfaces thus reducing the total number of the particles. The PWT-treatment continued to produce new suspended particles so that an increase in the total number of particles was observed for the particles smaller than 8 μm.

3.2.5 Results for Total Number of Particles Over Time

The efficacy of PWT depends on its ability to precipitate particles in bulk water so that particulate fouling could take place in a form of a soft sludge coating that can be cleaned by the shear force of flow [5, 14, 86]. This is so called a bulk precipitation mechanism. The size of particles in the bulk water is not as important as the overall number of suspended particles (even in submicron size) in the PWT-treatment method. This is because calcium ions in water have inverse solubility characteristics, which precipitate when heated at a heat transfer surface. But with the presence of suspended particles in water, these suspended particles actually provide additional surfaces to which the calcium ions can adhere thus reducing much of the precipitation fouling at the heat transfer surface.

Figure 3.6 presents the total number of particles for the no-treatment case versus the PWT-treatment case at flow rates of 500 and 800 l/h. It is of note that the total number of particles was the sum of all the particles from 1 μm to 280 μm. However, the number of particles over 8 μm in size was practically zero so that their contributions were negligible. Thus, in the present study, we can say that the total number of particles is the sum of small particles less than 8 μm. At the end of the test (i.e., at 66 h), there was an increase of 206 and 540 % for 500 and 800 l/h, respectively, in the total number of particles for the PWT-treatment as compared to the no-treatment case. The results obtained for the PWT-treatment cases at two flow rates showed similar trends with a drastic increase in the total number of suspended particles as compared to no-treatment cases. These results further support the bulk precipitation mechanism of the PWT.

This section investigated the mechanism of physical water treatment by counting the suspended particles in water. Hard water passed through a PWT device and water samples were collected at four different times during the test. The goal was to determine whether the PWT-treatment case produced more suspended particles than the no-treatment case. The no-treatment and PWT-treatment cases were tested at two flow rates of 500 l/h and 800 l/h, all at the room temperature (20–24 °C). The results showed that the total number of particles suspended in bulk fluid for the PWT-treatment case increased by 206–540 % depending on the flow rate. The results also showed that at the higher flow rate, which means more frequent exposure to PWT, one can have the higher precipitation of particles. The data presented here provide a support to the bulk precipitation/ electro-flocculation hypothesis of the PWT.

3.3 Fouling Experiments: Materials, Methods, Measurements and Characterization

This section describes the fouling test facilities and methods used at Chonbuk National University, Korea for the investigation of the efficacy of PWT using high frequency oscillating electric fields directly in water to mitigate calcium carbonate

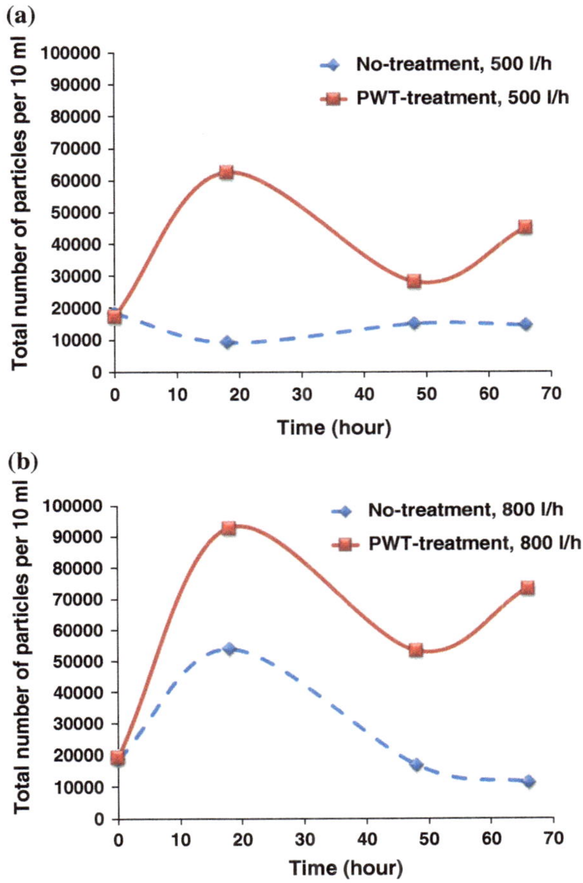

Fig. 3.6 Total number of particles over time at **a** 500 and **b** 800 l/h for no-treatment and PWT-treatment cases

fouling in a double-pipe heat exchanger. This study was limited to re-circulating experimental system where PWT is reported to be most effective especially in cooling-water applications. Artificially-made hard water was used to hasten the fouling process in a laboratory-scale experimental system.

3.3.1 Experimental Set-up

The schematic diagram of the fouling test experimental system is shown in Fig. 3.7. It was composed of a hard water tank, a centrifugal pump, a rotary flow meter, a two-graphite system to produce oscillating electric fields, a heat transfer test section, a chiller, a hot water circulating bath, LabVIEW device, and a personal computer. The hard water tank with a cover was made of plastic with a

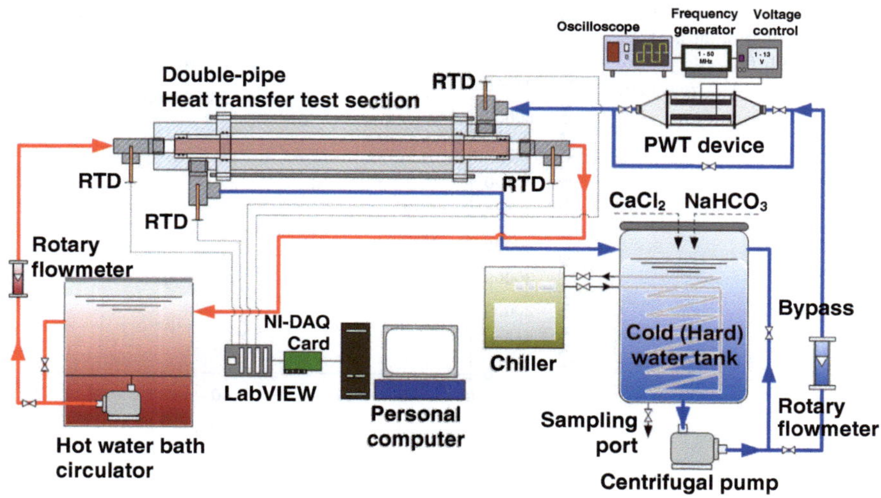

Fig. 3.7 Schematic diagram of the present fouling test experimental set-up. The whole experimental system including the heat transfer test section was grounded [14, 68]

maximum capacity of 200 l. Inside the tank was a copper coil connected to a chiller to maintain a constant temperature of the hard water during the tests. The centrifugal pump was controlled by an inverter to control the desired flow rate together with the rotary flow meter. The pump in the hot water re-circulating bath was also controlled by an inverter. The flow circulation loop was composed of copper tubes, expandable hoses, and heat-resistant Teflon tubes. A bypass flow was maintained throughout the tests to help keep the water temperature at the reservoir constant. Copper tube coils inside the hard water reservoir were connected to the chiller serving as the cooling heat exchange system. Two drain valves were positioned at the bottom of the reservoir tank for the purpose of both draining and sampling the hard water.

3.3.2 Heat Transfer Test Section

Figure 3.8 shows the heat transfer test section which was composed of copper tube, quartz crystal, Teflon heads, and fittings which were arranged in a double-tube, counter-flow configuration. The inner copper tube had dimensions of $1.385 \times 1.6 \times 49.9$ cm ($d_i \times d_o \times L$) and the quartz crystal had dimensions of $2.48 \times 2.81 \times 49.9$ cm ($d_i \times d_o \times L$). The cross-sectional areas for the copper tube and the annulus formed were 1.51×10^{-4} m^2 and 2.82×10^{-4} m^2, respectively. Both tubes were fitted in Teflon heads with connection fittings. The Teflon heads minimized the axial heat transfer losses in the test section. O-rings were

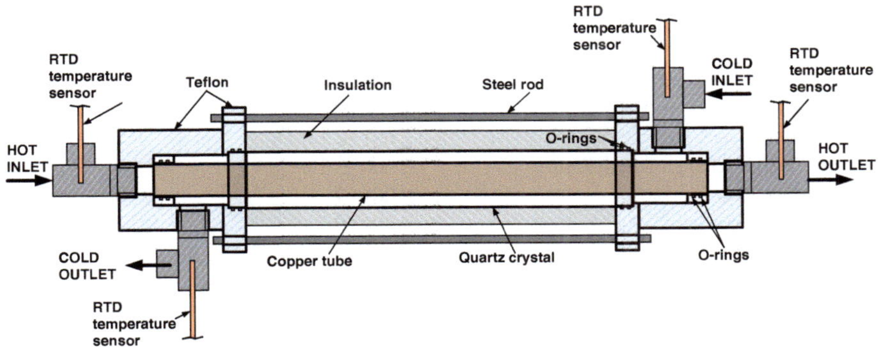

Fig. 3.8 The present double-pipe heat transfer test section for the fouling test experiments [88]

used on grooves at the inlet and outlet sides of Teflon head blocks for the copper tube to prevent any leakage between the hot water and the cold water. O-rings were also used between the quartz crystal and the Teflon head blocks to prevent leakage in the cold water side. Hot water flowed through the inner copper tube while cold water (i.e. hard water) flowed in opposite direction through the annulus gap between the two tubes. The hot water temperature was maintained constant by a hot water circulator bath. Static mixers were set in place at the outlet points of the hot and cold water sides. Four RTD temperature sensors (Sentech, Korea) were used to measure the inlet and outlet temperatures of both hot and cold water sides in the heat transfer test section. These RTDs were calibrated using a thermocouple bath calibration unit (Isothermal Technology Ltd), a precision thermometer (Model F250, Automatic Systems Laboratories), and the Measurement and Automation Explorer (MAX Version 4.0, National Instruments). The precision thermometer served as the reference temperature with a precision of ±0.01 °C. Four steel rods locked with nuts fixed the test section in place. Styrofoam was covered to the heat transfer test section to reduce any heat transfer losses to the surroundings. Table 3.4 gives the dimensions of the heat transfer test section.

3.3.3 Pre-Fouling Tests

Hot water was first passed through the inner tube of the test section for about 10–20 min to see any leakage in the inner tube. No water was flowing at the annulus at this stage. After making sure that there's no leakage in the inner tube, the cold water was then passed through the annulus of the test section, and likewise was checked for any leakages. When no leakage was found, both hot and cold water sides were kept flowing in opposite directions in the heat transfer test section for at least one hour to stabilize them at the desired temperatures and flow rates. Distilled water was used. The stabilized system at the desired parameters was

Table 3.4 Heat transfer test section details

Unit	Copper tube	Quartz crystal
Inner diameter (d_i) (mm)	13.85 ± 0.05	24.80 ± 0.05
Outer diameter (d_o) (mm)	16.00 ± 0.05	28.10 ± 0.05
Length (mm)	631 ± 1.0	499 ± 1.0
Effective heat exchange length (mm)	600 ± 1.0	–
Inner surface area (cm^2)	261.07	–
Outer surface area (cm^2)	301.59	–
Mean surface area (cm^2)	281.33	–
Area (cross sectional) (cm^2)	1.51	–
Area (annulus) (cm^2)	–	2.82

continuously run to test for the discrepancy of heat transfer rates between the hot and cold water sides. The heat transfer test section was covered with a Styrofoam insulation to minimize any heat loss to the surroundings. The RTDs were also covered with Styrofoam to prevent any heat transfer from the surroundings that might affect the temperature readings. Both heat transfer test section and the entire fouling set-up were grounded to lessen the possible effect of electrical noise and static electricity on the temperature readings and in the process as a whole [89].

3.3.4 Preparation of Artificial Hard Water

One of the most common forms of scale is calcium carbonate, which possesses inverse solubility characteristics (i.e., its solubility decreases when temperature increases) [1]. In the present study, we wanted to have calcium carbonate fouling so that we prepared hard water artificially for more controlled water chemistry by mixing chemical reagents of calcium chloride ($CaCl_2$) and sodium bicarbonate ($NaHCO_3$) in distilled water. The hardness of water in the study varied from 350 to 550 mg/l. The desired water hardness was obtained by mixing two chemicals at right proportions (see Table 3.5): Calcium chloride ($CaCl_2$) (Showa Chemical Co., Ltd., Japan) and Sodium bicarbonate ($NaHCO_3$) (Showa Chemical Co., Ltd., Japan). The mixture of both chemicals resulted in the following chemical reaction:

$$CaCl_2 + 2NaHCO_3 \rightarrow CaCO_3 + 2NaCl + H_2O + CO_2 \qquad (3.3.1)$$

During the tests, there were no additional chemicals (i.e., $CaCl_2$ or $NaHCO_3$) added into the system so that we allowed the water hardness level to gradually go down due to fouling formation. The initial water hardness was at 350 ± 20 and 550 ± 20 mg/l for all tests. Table 3.6 gives the approximate classifications of water hardness. Thus, the present water hardness used is considered to be very hard so as to hasten the fouling process.

The hard water reservoir was first filled with distilled water at a volume of approximately 150 l after cleaning. With the use of a by-pass line, the distilled

Table 3.5 Preparation of artificial hard water [14, 68]

Distilled water volume: 0.150 m^3	Hardness (mg/l)	
	350 mg/l	550 mg/l
Calcium chloride ($CaCL_2$: Mw = 110.98 g/mol)	58 g	97 g
Sodium bicarbonate ($NaHCO_3$: Mw = 84.01 g/mol)	88 g	147 g

Table 3.6 Classification of water hardness [34]

Water hardness (mg/l)	Classification
Less than 17.1	Soft
17.1–60	Slightly hard
60–120	Moderately hard
120–180	Hard
180 and over	Very hard

water in the reservoir was continuously circulated (not through the whole fouling system set-up, but only in the reservoir). $CaCl_2$ powder was added to the reservoir with gentle stirring using a metal stick and was left for 10 min to dissolve. Subsequently, $NaHCO_3$ powder was also added to the reservoir and was gently stirred. After 5–7 min, about 100-ml of water sample was taken out from the bottom of the reservoir (see Fig. 3.7) and used for water chemistry measurements. After 10 min of mixing, two more 100-ml hard water samples were collected for water chemistry measurements and used as the initial water chemistry data (t = 0). Samples were taken three more times at 12–18 h interval. Each water sample collected was about 100 ml.

3.3.5 Water Analyses

The water chemistry of the circulating water was analyzed during the scaling process by taking water samples at a volume of 100 ml at different time intervals from the sampling port located at the bottom of the hard water tank (see Fig. 3.7). The water was checked for hardness, electrical conductivity and pH from t = 0 until t = 36–40 h. Table 3.7 presents the methods used for water chemistry analysis and their calibration procedures. The hardness of water primarily refers to the existence of the dissolved minerals of calcium and magnesium and other metals in water [36]. These minerals were dissolved from soils and rocks in underground water. Alkalinity relates to the ability of water to buffer pH. The major ions that gives the total alkalinity of water includes hydroxide ion (OH^-), carbonate ion (CO_3^-), and bicarbonate ion (HCO_3^-) . Electrical conductivity of water pertains to the ability of water to conduct electricity depending on the concentration and mobility of ions, the oxidation state and the temperature of

Table 3.7 Water analyses method and calibration

Parameter	Method/device	Details	Calibration
Hardness (mg/l)	Ethylenediaminetetraacetic acid (EDTA) titration	HI 3812, Hanna Instruments, Hungary	Manufacturer's calibration
Alkalinity (mg/l)	Hydrochloric (HCl) titrimetric method	HI 3811, Hanna Instruments, Hungary	Manufacturer's calibration
Electrical conductivity (μS/cm)	Portable electrical conductivity meter	Cond 340i, WTW, Germany	Using a 0.01 mol/L KCl control standard
pH	Portable pH meter	HI 8424, Hanna Instruments, Portugal	Using buffer solutions at pH = 7.0 and pH = 10.0

Table 3.8 Initial water chemistries for distilled water, tap water, and artificial hard water used in the present study

Parameter	Distilled water	Tap water	Artificial hard water	
			350 mg/l	550 mg/l
Hardness (mg/l)	3 ± 1	42 ± 3	350 ± 20	550 ± 20
Alkalinity (mg/l)	9 ± 1	33 ± 2	410 ± 30	630 ± 10
Electrical conductivity (μS/cm)	10 ± 1	210 ± 2	3480 ± 20	5430 ± 20
TDS (mg/l)	7 ± 1	134 ± 2	2230 ± 7	3470 ± 13
pH	5.6 ± 0.3	6.3 ± 0.3	8.0 ± 0.02	7.55 ± 0.05

water. The effect of temperature is significant on water conductivity. Generally, the conductivity of water increases with the increase in temperature. Table 3.8 shows the initial water chemistry of distilled water, tap water, and artificial hard water used in the present study.

3.3.6 Fouling Tests

The inlet temperatures of the hot and cold water sides of the heat transfer test section were maintained constant throughout the tests. Hot and cold-water flow through the double-pipe heat exchanger in opposite directions at constant flow rates. The hot water flows through the inner copper tube while the cold-water (i.e., hard water) flows through the annulus gap formed between the two concentric tubes. The mean temperatures for both hot and cold water sides served as the reference temperatures for specific heat and density values. The temperature readings were measured every five minutes with a real-time DAS using LabView program (National Instruments). Both the heat transfer test section and the PWT device were electrically grounded. The duration for each fouling test was about 30–40 h.

After each fouling test, the artificial hard water was drained out of the system loop. Since the hard water was drained right away after the test, the remaining scales in the hard water reservoir as well as in the system loop could be relatively easily removed. We cleaned the reservoir tank with a brush and flushed it with high pressure water jet. Then, tap water was recirculated through the loop continuously for 2–5 h. At this point, the system was considered to be ready for the next test and new distilled water of 0.150 m^3 (150 l) was poured into the reservoir. A new, smooth and clean copper tube was installed in the heat transfer test section for each fouling test. Prior to this cleaning process, the heat transfer test section was removed from the system loop and the scaled pipe was dried off for inspection and SEM photography.

The heat transfer rate from the hot-water side was estimated using the following equation:

$$Q_h = \dot{m}_c c_{ph} \Delta T_h \tag{3.3.2}$$

The heat transfer rate at the cold-water side can be described in a similar manner:

$$Q_c = \dot{m}_c c_{pc} \Delta T_c \tag{3.3.3}$$

The discrepancy in heat transfer rates between the hot and cold-water sides was found to be ±8 %. The heat transfer rate in the cold-water side Q_c was used to calculate the overall heat transfer coefficient considering heat losses in the hot-water side might have incurred to the surroundings although the hot-water side was insulated.

The fouling resistance was calculated using the following equation [5, 6, 14, 18]:

$$R_f = \frac{1}{U_f} - \frac{1}{U_{ini}} \tag{3.3.4}$$

where U_f is the overall heat transfer coefficient for the fouled case, while U_{ini} is the overall heat transfer coefficient for the initial clean condition. This overall heat transfer coefficient was calculated using the following equation [5, 6, 14, 18]:

$$U = \frac{Q_c}{A_o \Delta T_{lmtd}} \tag{3.3.5}$$

where ΔT_{lmtd} is the log-mean-temperature difference which was determined from measured temperatures at the inlet and outlet of hot and cold water which is given as [5, 6, 14, 18]:

$$\Delta T_{lmtd} = \frac{\left(T_{h,out} - T_{c,in}\right) - \left(T_{h,in} - T_{c,out}\right)}{\ln\left[\frac{\left(T_{h,out} - T_{c,in}\right)}{\left(T_{h,in} - T_{c,out}\right)}\right]} \tag{3.3.6}$$

The surface area used to calculate the overall heat transfer coefficient was based on the outer diameter of the inner copper tube and an effective heat transfer length of 0.6 m. The ΔT_{lmtd} was maintained around 56–58 °C in all tests. According to the method of Kline and McClintock [90, 91] the uncertainties of fouling resistance and overall heat transfer coefficient were 15–18 % and 6–8 %, respectively.

3.3.7 Cleaning Procedure

After each test, the hard water tank was cleaned thoroughly. After the previously used artificial hard water was completely drained, a tap water jet using a hose was then flushed onto the inside wall of tank together with a cleaning brush to remove any scales formed inside the tank. Traces of $CaCO_3$ agglomerated deposits were drained out of the tank. After this, tap water was introduced into the tank and was circulated in the system without the test section for at least one hour. When most visible traces of the hard water previously used were gone, the tap water was completely drained out and new distilled water up to 0.150 m^3 (150 l) was supplied to the tank. No chemicals were added during the cleaning process. The RTDs were thoroughly cleaned by using a fine grain sandpaper to scratch off gently the scales formed on the RTD surface, and then wiped with tissue paper to ensure a clean RTD surface for the next test. This was done so as to lessen the effect of scales on the RTD which might have produced erroneous temperature measurement. After completion of each test, the test section was disassembled, and the fouled tube was dried on a display board. The quartz crystal tube was thoroughly cleaned with tap water and wet tissue to remove the scales in the inside of the quartz tube. The inside of the Teflon heads were also cleaned and dried. After which, a new copper tube was installed in the test section, and the quartz tube was also installed putting Teflon seal on both ends of the two tubes to ensure no leakage, as mentioned early.

3.3.8 Analysis of $CaCO_3$ Scales

3.3.8.1 Scanning Electron Microscopy

Deposit samples of approximately 0.5 × 0.5 cm were obtained from the fouled tubes (both no-treatment and PWT-treatment cases) by manually cutting the tube using a saw blade. SEM images were taken at a magnification of 5,000× to examine the deposit surface topography and geometry. The samples for SEM were prepared carefully making sure that there were no scratches on their surfaces. Care was given not to contaminate the deposit with any impurities during the sample preparation. The deposit samples were coated by gold through low-vacuum sputter coating in order to prevent the accumulation of static electric charge during the irradiation of electron.

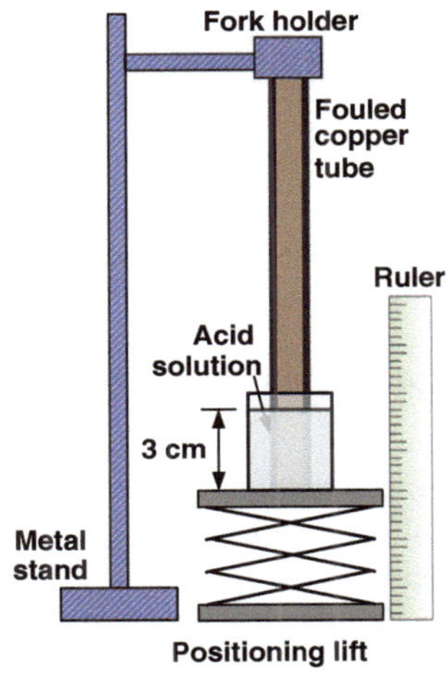

Fig. 3.9 Schematic drawing of the present set-up of the test for deposited calcium carbonate scales. Hydrochloric acid solution was used to dissolve the scales [88]

3.3.8.2 X-ray Diffraction (XRD)

XRD analyses were conducted on 1 × 1 cm samples taken from the fouled tubes by manual cutting to determine and characterize the crystallographic structure of the deposits for both the no-treatment and PWT-treatment cases. The $CaCO_3$ scales on the fouled tube were expected to exhibit calcite, aragonite, or vaterite crystal patterns of $CaCO_3$ or their combinations.

3.3.8.3 Calcium Ion Concentration Analysis

After drying the $CaCO_3$ scale deposits on the copper tube at the end of each test, the amount of calcium ions in the deposited $CaCO_3$ scale samples was measured using an inductively coupled plasma mass spectrometer (ICP-MS, 7500a, Agilent, USA). Due to the brittleness of some scales, it was hard to cut the exact same length of scaled tubes without detaching some scales from the tubes, so that the following method was used. The $CaCO_3$ scales still attached to the tube ($L = 3$ cm) were dissolved in a 75-ml hydrochloric acid (HCl) solution. Figure 3.9 shows the present schematic drawing of the test set-up for dissolution of scales for analysis by ICP-MS. Though the thickness of the scales varied through the length of the tube, a representative length of 3 cm of scaled copper tube with fouling deposit was checked. The solution was prepared by 70/30 % volume ratio of distilled water to HCl acid (HCl 35–37 % Assay). The scales took only 30–60 s to dissolve. The solution was then divided into seven samples at 10 ml each from which the standard deviation was calculated.

Table 3.9 The experimental conditions and parameters in the present study

Parameter	Hot water	Cold (hard) water
Inlet temperature (°C)	85.5 ± 0.5	23.5 ± 0.5
Flow velocity (m/s)	1.1	0.3
Mass flowrate (kg/s)	0.162	0.084
Water hardness (mg/l)	10–50	550

3.4 Effect of Varying Frequency on Fouling Mitigation

In this section, the effect of varying the frequency of the electric field at a fixed voltage on the mitigation of scaling of a double-pipe heat exchanger was investigated. It is speculated that at higher frequency, there is a higher tendency for ions to collide that could lead to the formation of colloidal particles in bulk water. When particles are precipitated at the bulk water instead of directly to the heat transfer surface, it is believed to be easier to remove by shear forces from fluid flow, thus mitigating the fouling.

3.4.1 Experimental Set-up and Method

The fouling test system (see Fig. 3.7), the double-pipe heat exchanger (see Fig. 3.8), and the experimental methods and tests are all described in the previous section. The test conditions for the present study are shown in Table 3.9. The inlet temperatures of the hot and cold water sides of the heat transfer test section were maintained throughout the tests at 85.5 ± 0.5 °C and 23.5 ± 0.5 °C, respectively. The inlet-outlet temperature difference in the hot water side was maintained at 2–4 °C to provide uniform fouling conditions along the test section. The cold water flow rate was maintained at 8.5×10^{-5} m^3/s which was equivalent to a flow velocity of 0.3 m/s at the heat transfer test section (cold water side). The hot water side had a constant flow velocity of 1.1 m/s (1.67×10^{-4} m^3/s). The mass flow rates for the hot and cold water sides were maintained constant at 0.162 and 0.084 kg/s, respectively. The temperature readings were measured every five minutes with a real-time DAS using LabVIEW program (National Instruments). Both the heat transfer test section and the PWT device were electrically grounded. The duration for each fouling test was about 30–36 h. After each test, the experimental system was thoroughly cleaned using the cleaning procedures described in the previous section.

The present fouling tests were conducted in two cases: a) reference (no-treatment) case, and b) PWT-treatment case. The reference case and the PWT-treatment cases were compared to view any effects of PWT in the fouling resistances and water chemistry through time. The preparation and cleaning procedures, water chemistry, experimental conditions such as inlet temperatures and velocities, and the equipment used were the same for all tests. A fixed voltage of ±10 V was used at four different frequencies for the PWT-treatment cases: 13.56, 1 MHz, 100, and 1 kHz. The fouling resistance was calculated, water analysis was conducted, and SEM and photographic images of the scale deposits were obtained.

3.4.2 Experimental Results and Discussion

3.4.2.1 Fouling Resistance

Figure 3.10 shows the fouling resistances versus time for both no-treatment and PWT-treatment cases. All fouling resistances for both cases did not show any induction period. An induction period is the time wherein a stable crystal nucleation takes place and slowly spreading out laterally until completely covering the heat transfer surface with scale deposits [92, 93]. It is indicated by a straight line in the beginning of the fouling test. The high water hardness, high heat transfer surface temperature, and low flow velocity must have affected the negligible induction periods . The presence of vast number of calcium ions in bulk water at a low flow velocity (i.e., low flow shear rate) should have been the primary factor in the rapid scale deposition on the first few hours of the test as indicated in the drastic increase of fouling resistance. The fouling resistance for the no-treatment case increased exponentially to its maximum value of 4.7×10^{-4} m^{2}K/W after t = 12 h of test and then gradually decreased and reached 4.2×10^{-4} m^{2}K/W at the end of t = 36 h. The whole copper tube surface was already fully covered with white scale deposits after t = 6 h when checked visually. At the end of the test, the dried scale deposits on the copper tube showed rough deposit surfaces (see Fig. 3.11b).

The PWT-treatment cases showed lower fouling resistances as compared to the no-treatment case. In the case of PWT-treatment at f = 13.56 MHz, the fouling resistance decreased by 60 % from the no-treatment case at the end of the test. It reached its asymptotic value of 1.65×10^{-4} m^{2}K/W at t = 34 h. The fouling deposit layer on the copper showed a smoother surface (Fig. 3.11c) as compared to the no-treatment case. It was obviously thinner when checked with visual inspection though no thickness measurement was done. The scale deposits had a bluish-green color as opposed to the white-colored scales at the no-treatment case. For the case of PWT-treatment at f = 1 MHz, the fouling resistance had an asymptotic value of 2.7×10^{-4} m^{2}K/W after t = 36 h, which was a 36 % decrease from the no-treatment case. The fouling deposits were white in color and with rough surface though lesser in the degree of roughness than that of the no-treatment case (see Fig. 3.11d). At a much lower frequencies of 100 and 1 kHz, the fouling resistances decreased by about 25 and 16 %, respectively from the no-treatment case after t = 36 h. Both fouling deposits had white-colored scales (Fig. 3.11e and f). As one can notice in the fouling resistances, the first 10–12 h of test is crucial in the crystal growth and adhesion for crystallization fouling using

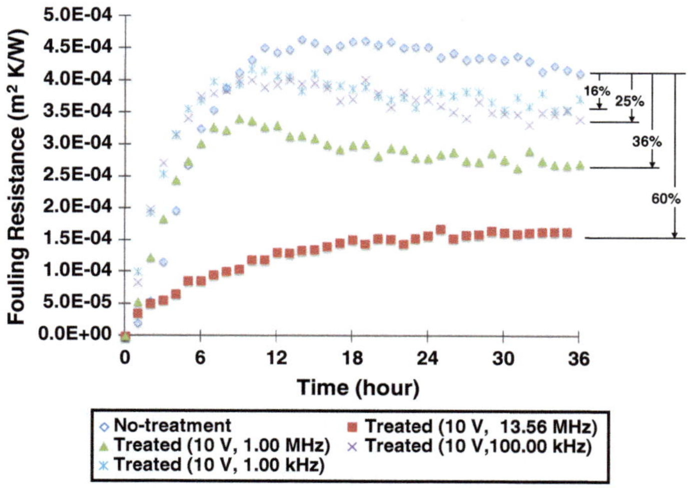

Fig. 3.10 Fouling resistance versus time for both no-treatment and PWT-treatment cases at a fixed voltage of 10 V and at varying frequencies (0.3 m/s cold-water side velocity) [14]

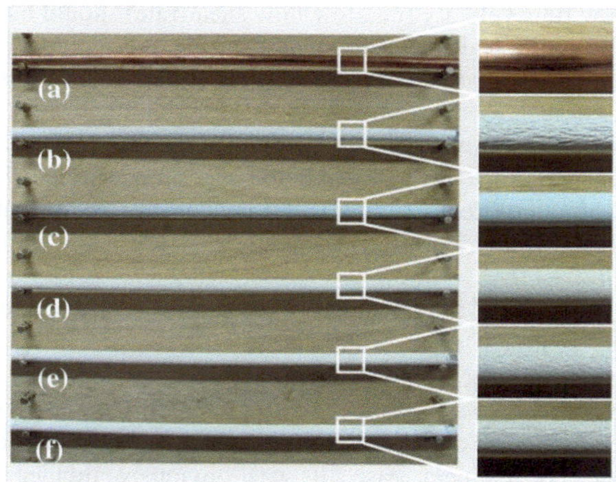

Fig. 3.11 Photographic images of the **a** new and clean tube **b** no-treatment case, and the fouled tubes at different PWT-treatment conditions: **c** 10 V, 13.56 MHz **d** 10 V, 1 MHz **e** 10 V, 100 kHz, and **f** 10 V, 1 kHz. The enlarged photos are shown at the *right side* for the areas in *white square* [14]

artificial hard water. The effect of frequency on the mitigation of fouling in the present study had better performance as frequency was increased. After each test, the surfaces of the graphite electrode plates did not show any scaling.

3.4.2.2 SEM Images

The SEM photographs for both no-treatment and treated cases are shown in Fig. 3.12. The specimens with fouling deposit for SEM photography were prepared by drying the fouled tubes thoroughly after each test at room temperature.

Fig. 3.12 SEM photographs of the fouled tubes for no-treatment and PWT-treatment cases. A 5 × 5 mm specimen was cut from the fouled tube after each fouling test for SEM imaging [14]

An approximately 5 by 5 mm specimen (i.e. copper tube with scale on surface) was cut from the fouled tube using a cutting saw. The SEM results in Fig. 3.12 show different crystal structures between the no-treatment and the treated cases. The no-treatment case (Fig. 3.12a) showed sharp and needle-like crystals. The pointed structures were less than 1 μm in size. The PWT-treatment case at f = 13.56 MHz (Fig. 3.12b) showed blunt crystal tip and edges and were a bit thicker in size (more than 1 μm). The PWT-treatment-cases at f = 1 MHz, at f = 100 kHz, and at f = 1 kHz (Fig. 3.12c, d, and e, respectively) showed flower-like crystal structures and agglomerated scales. The pointed form of scale at the no-treatment case could be more adherent to the heated surface as compared with the non-pointed structures in the PWT-treatment case. Aragonite crystal structures are known to be harder to remove as compared to the calcite crystal structures. Thus, the PWT-treatment results suggest an easier to remove scales.

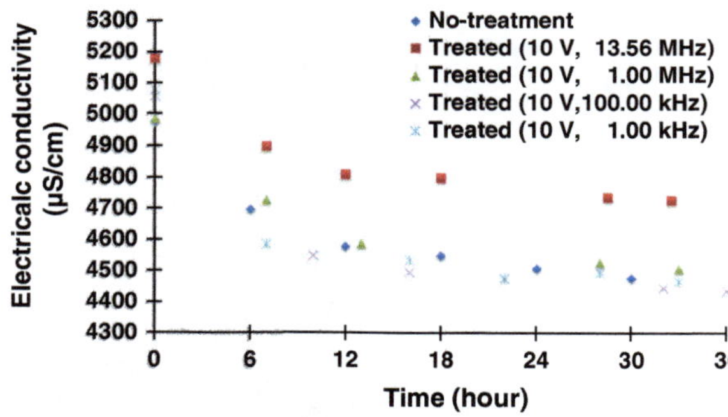

Fig. 3.13 Electrical conductivity measurements at different times for no-treatment and at PWT-treatment cases at initial water hardness of 550 mg/l and a flow velocity of 0.3 m/s [14]

3.4.2.3 Water Analysis

The water properties in the fouling tests was also measured and analyzed. The water analyses were done within an hour from water sampling. It should be noted that the initial values of water quality for all tests were made sure to be similar, and no chemical reagents were added once each test was started. The electrical conductivity, pH, and total hardness of water (refer to Figs. 3.13, 3.14 and 3.15) were analyzed every 6–10 h with a sample volume of 100–200 cm^3.

Electrical Conductivity

Figure 3.13 shows the electrical conductivity through time of both no-treatment and PWT-treatment cases. The initial conditions were kept similar at 5,000–5,200 μS/cm. The results showed a similar decreasing trend for all tests. The no-treatment case dropped by 10 % after t = 30 h from its initial value. The electrical conductivities for the PWT-treatment cases also similarly dropped from 8.6–12.3 % depending on the frequency applied. The electrical conductivity is directly related to the number of conductive ions in the solution, thus with the continuous deposition of $CaCO_3$ scales, electrical conductivity decreased through time.

pH

The acidity or alkalinity of a solution says a lot in its precipitating capability. In the present study, the pH of the hard water was measured from the similar initial value for all cases, until t = 36 h. The results shown in Fig. 3.14 signifies that as mineral crystals are depositing on heat transfer surfaces, the alkalinity of the solution decreases. For the case of the PWT-treatment case at f = 13.56 MHz, there was 4 % decrease in pH after 32 h as compared with 9 % decrease for the no-treatment case. The results for the other PWT-treatment cases varied from 6–8 % drop in pH from their initial values.

Total Hardness

In the present study, distilled water with negligible water hardness was used to mix the solution for producing artificial hard water at 550 mg/l. Thus, the main component of water hardness was calcium ions. Figure 3.15 shows the water

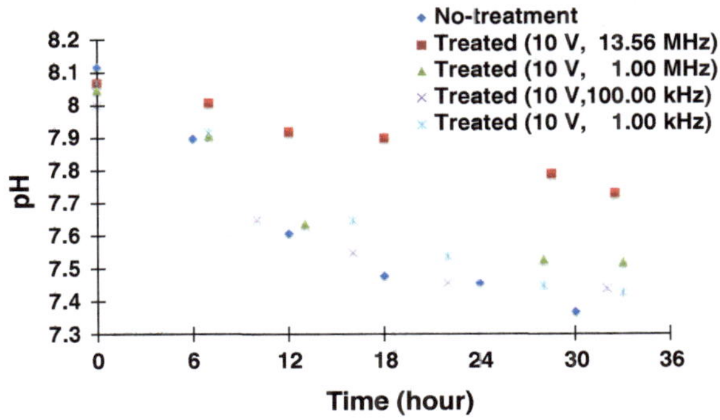

Fig. 3.14 pH measurements at different times for no-treatment and at PWT-treatment cases at initial water hardness of 550 mg/l and a flow velocity of 0.3 m/s [14]

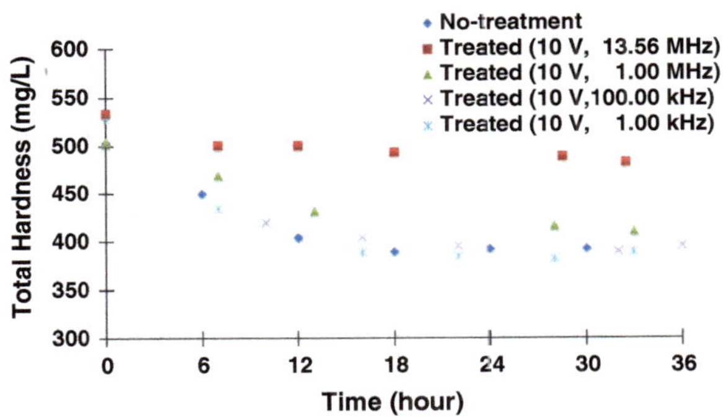

Fig. 3.15 Total water hardness measurements at different times for no-treatment and at PWT-treatment cases at initial water hardness of 550 mg/l and a flow velocity of 0.3 m/s [14]

hardness with respect to time in the present study. The water hardness at the no-treatment case decreased by 22 % after the fouling test. The PWT-treatment case at f = 13.56 MHz showed the lowest percentage decrease at 9.6 %, and as the frequency was lowered, the percentage decrease also increased to 18.6, 21 and 26 % for f = 1 MHz, f = 100 kHz and f = 1 kHz, respectively. Calcium ions combine with carbonate ions found in water to form the scales so that the more deposition that takes place, the lesser will be the water hardness through time.

In the present study, the PWT used oscillating electric fields which were generated using two graphite plate electrodes submerged in water. The PWT device was operated at 10 V and in a frequency range of 1 kHz–13.56 MHz. Artificial hard water of 550 mg/l of $CaCO_3$ hardness was utilized in all fouling tests, with cold water flow velocity of 0.3 m/s. The present results showed that the PWT device performed better with increasing frequency. In the present study the maximum frequency tested was 13.56 MHz, where the best performance was obtained for the mitigation of mineral fouling. At this frequency, the fouling resistance decreased by 60 % from the no-treatment case. The SEM photograph also at this frequency depicted a calcite form of scale which is easier to remove with enough flow shear forces. One may obtain a better performance at a frequency greater than 13.56 MHz. However, since 13.56 MHz is the frequency allowed for industrial applications by Federal Communications Commission (FCC), it may be concluded that the present test showed the maximum performance of the present PWT operation.

3.5 Effect of Varying Flow Velocity and Initial $CaCO_3$ Concentration

The efficacy of PWT especially in laboratory experiments depends on many parameters and conditions which includes the flow velocity and $CaCO_3$ concentration. According to many studies [5, 6, 10, 18, 28, 46], the effectiveness of PWT depends on the shear forces created by fluid flow. In this section, the effect of varying flow velocity on the mitigation of scaling was investigated. Furthermore, two high initial hard water concentrations were used to hasten the fouling process and were checked on their effect on fouling formation and the mitigation by PWT.

3.5.1 Experimental Set-up and Method

Chapter 3 describes the experimental system (see Fig. 3.7) and the heat transfer test section (see Fig. 3.8) used in the present tests. The hot and cold water-side temperatures at the heat transfer test section were maintained constant using a hot water re-circulating bath and a chiller, respectively. Hot water flowed inside the copper tube at a constant velocity, while the cold water made of artificial hard water passed through the annulus gap flowing in the opposite direction at 3 different velocities (0.3, 0.8, and 1.0 m/s). The corresponding velocity at the graphite electrode plates was 0.12, 0.32, and 0.40 m/s, respectively. The heat transfer rate varied from 2.9 to 5.8 kW depending on the velocity at the cold-water side. RTDs connected to a real-time data acquisition system (LabVIEW 7.0) were used to measure temperatures every 5 min for a test period of 36–40 h. Table 3.10 shows the summary of the experimental parameters.

The present study was divided into two parts: The first set of tests was done at a high artificial water hardness of 550 mg/l to investigate the feasibility of PWT method using oscillating electric fields to mitigate mineral fouling, and to further determine the optimum voltage and frequency of the electric field.

The second set of tests utilized the optimum values for voltage and frequency to generate oscillating electric fields at a lowered artificial water hardness of 350 mg/l and at three different cold water-side flow velocities at the heat exchanger (0.3, 0.8 and 1.0 m/s). All experimental procedures, materials, equipment, inlet temperatures for hot and cold water sides, and volume of water used were the same for all tests.

3.5.2 Results and Discussion

3.5.2.1 Water Analysis

In the present study, the pH, calcium carbonate hardness, and electrical conductivity of the hard water medium during the fouling test was monitored and

Table 3.10 Summary of the experimental parameters

Parameter	Hot water	Cold (hard) water
Inlet temperature (°C)	85.5 ± 0.5	23.5 ± 0.5
LMTD (°C)	53–56	
Flow velocity (m/s)	1.1–1.3	0.3–1.0
Mass flowrate (kg/s)	0.162–0.194	0.084–0.281
Heat transfer rate (W)	2,900–5,800	
Reynolds number	43,000–53,000	1,100–3,900
Water hardness (mg/l)	10–50	350–550

checked. There were no additional chemicals added during the tests, so that the water properties were not maintained constant throughout the test; only the initial water condition was made sure to be the same for all tests.

The acidity or alkalinity of a solution portends its precipitating capability. When the pH of water increases, the conversion of bicarbonate to carbonate also increases, so that it has a higher potential to form calcium carbonate. The pH of water gives an indication of the scaling ability of water. At high pH or more alkaline, there is more chances of higher precipitation to occur. Parsons et al. [94] studied the effect of controlling pH on the calcium carbonate scale formation. They found out that controlling the pH of the scaling solution leads to non-effective physical water conditioning because controlling pH directly controls the scaling rate. Andritsos and Karabelas [95] reported that when pH was increased from 8.8 to 10, the deposition of calcium carbonate also increased from 2 to 12 mg/cm^2, a 500 % increased only after a 2-h run. Dawson [96] similarly observed a 500 % increase in scale precipitation when the pH was changed from 7 to 8. The pH of water was measured at different intervals during the entire fouling test for the no-treatment and PWT-treatment cases. Figure 3.16 shows the results of pH measurements for water hardness of 550 (Fig. 3.16a) and 350 mg/l (Fig. 3.16b). Both water hardness tests had initial pH of 7.5–8.0. Higher pH values of water signify higher precipitating ability [97]. In the present study, it was expected that the pH would drop as the fouling process progressed primarily due to the deposition of calcium carbonate on the hot copper tube surface. As the mineral crystals are depositing on the hot solid surfaces, the pH of the solution decreases. The more calcium carbonate scale deposition, the lower the pH of water becomes. For the PWT-treatment cases, there was a 2–6 % decrease in pH after 36–40 h as compared with a 7 % decrease for the no-treatment case depending on the initial water hardness and flow velocity. There was not much big difference in the pH results for no-treatment and PWT-treatment cases because of the very high initial water hardness which basically maintained its being alkaline, above pH 7.0, throughout the 36–40 h test. Of note is that no additional chemicals were added during the fouling process. The results showed lowest pH value at the no-treatment case for both water hardness conditions at the end of the test.

In the present study, distilled water was used to mix the solution for producing artificial hard water at 550 and 350 mg/l. Thus, the main component of water

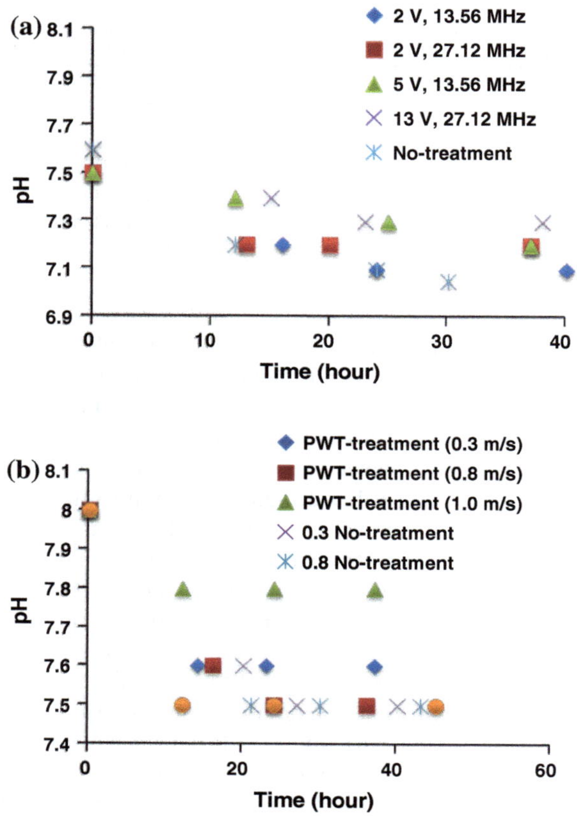

Fig. 3.16 pH measurements at different times for no-treatment and PWT-treatment cases at initial water hardness of **a** 550 mg/l and **b** 350 mg/l [88]

hardness was calcium ions. Figure 3.17a and b show the water hardness measurements for 550 and 350 mg/l, respectively. In Fig. 3.17a, the no-treatment case decreased by 40 % at the end of the test. The PWT-treatment cases decreased also by about 39 % from their initial value depending on the treatment condition. At water hardness of 350 mg/l (see Fig. 3.17b), the no-treatment and PWT-treatment cases dropped from their initial values by 15–45 % and 13–22 %, respectively. The largest drop was at no-treatment case and at a velocity of 0.3 m/s, where the highest fouling thermal resistance was obtained. Calcium ions combine with carbonate ions found in water to form the scales so that the more deposition that takes place, the lesser will be the water hardness through time.

Figure 3.18 shows the electrical conductivity through time of both no-treatment and PWT-treatment cases. The initial conditions were kept similar at 5,430 ± 20 and 3,480 ± 20 μS/cm (see Figs. 3.18a and b, respectively). The results showed a similar decreasing trend for all tests. The no-treatment case dropped by 13–16 % after t = 30–37 h from its initial value and the PWT-treatment cases also similarly dropped from 8–12 % depending on the frequency applied and water hardness.

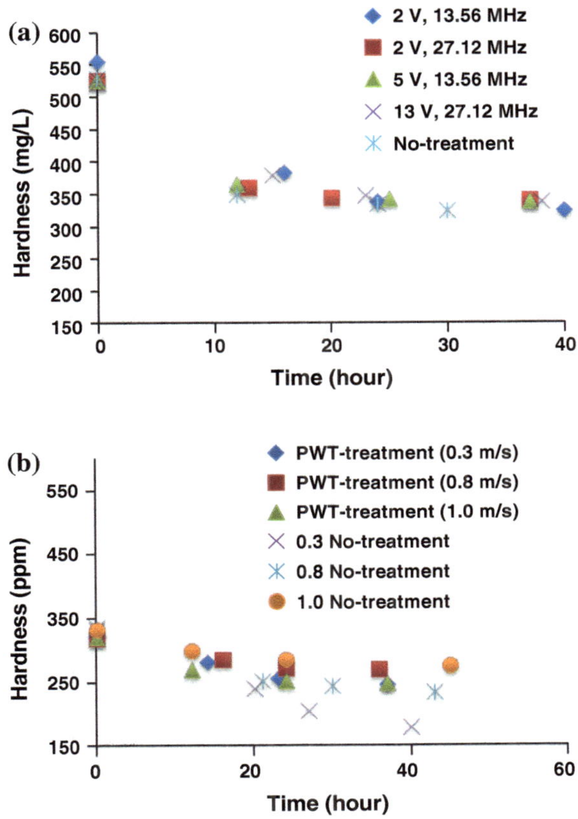

Fig. 3.17 Water hardness measurements at different times for no-treatment and PWT-treatment cases at initial water hardness of **a** 550 mg/l and **b** 350 mg/l [88]

The electrical conductivity is directly related to the number of conductive ions in the solution, thus with the continuous deposition of $CaCO_3$ scales, electrical conductivity decreased through time.

3.5.2.2 Calcium Ion Concentration on Scale Deposits

Figures 3.19 and 3.20 show the results of the chemical analyses through ICP-MS of the scale deposits for the no-treatment and PWT-treatment cases at 550 and 350 mg/l, respectively. In Fig. 3.19, the no-treatment case had the most amount of calcium as compared with the PWT-treatment cases. The calcium concentration dropped from 4–46 % at the PWT-treatment case as compared with the no-treatment case. This signifies a lower calcium carbonate deposition for the PWT-treatment case. Considering both no-treatment cases (Figs. 3.19 and 3.20) at velocity of 0.3 m/s, the deposit at 350 mg/l hardness showed 25 % lower calcium concentration than the deposit at 550 mg/l hardness, indicating good consistency

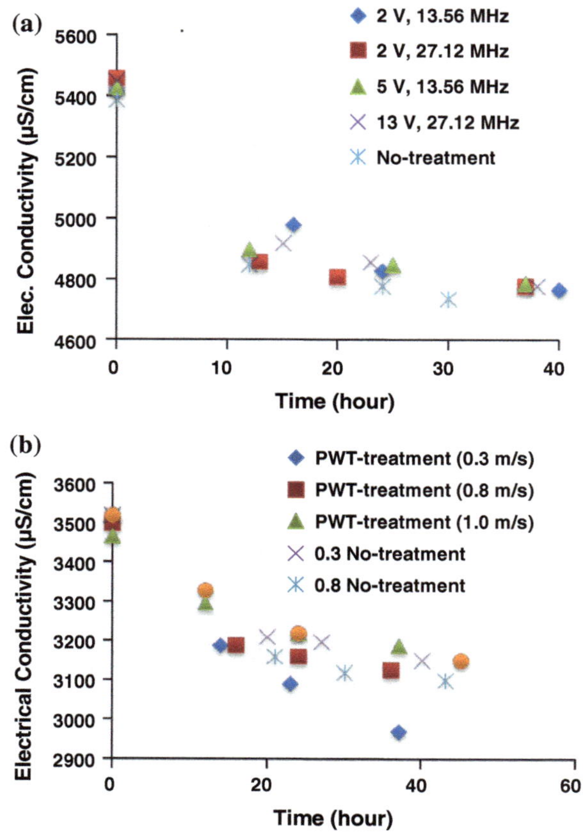

Fig. 3.18 Electrical conductivity measurements at different times for no-treatment and PWT-treatment cases at initial water hardness of **a** 550 mg/l and **b** 350 mg/l [88]

of results, i.e., more deposition at higher hardness. Figure 3.20 also shows a similar trend for calcium concentration where higher calcium concentrations were obtained at lower velocities for both no-treatment and PWT-treatment cases.

The calcium concentrations at the no-treatment cases were also consistently higher than those with the PWT-treatment for each test velocity. These results corresponded well with the fouling thermal resistance results wherein the no-treatment case had the highest fouling thermal resistance values (see Figs. 3.21 and 3.24). Chibowski et al. [98] found that the surface properties, materials, and temperature could affect the amount of deposited calcium carbonate in quiescent conditions. But generally, their use of magnetic field reduced the firmly adhered $CaCO_3$. The results in the present study similarly suggest that the use of high-frequency electric fields generally reduced the amount of calcium depositing on the surface. It is of note though that through visual inspection, $CaCO_3$ deposition was uneven throughout the length of the copper tube surface.

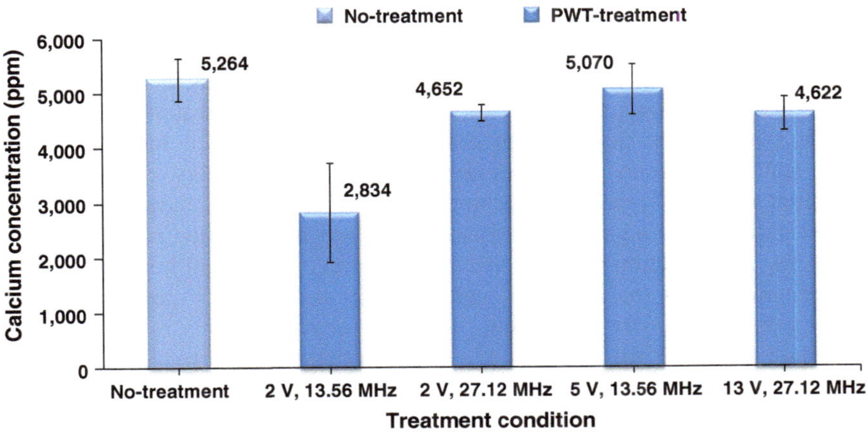

Fig. 3.19 Calcium concentration of the fouling deposits using ICP-MS for no-treatment and PWT-treatment cases (550 mg/l and 0.3 m/s) [88]

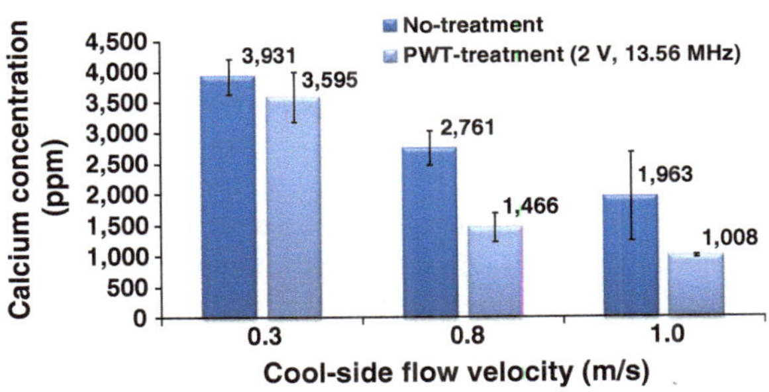

Fig. 3.20 Calcium concentration of the fouling deposits using ICP-MS for no-treatment and PWT-treatment cases at various flow velocities from 0.3 to 1.0 m/s (350 mg/l) [88]

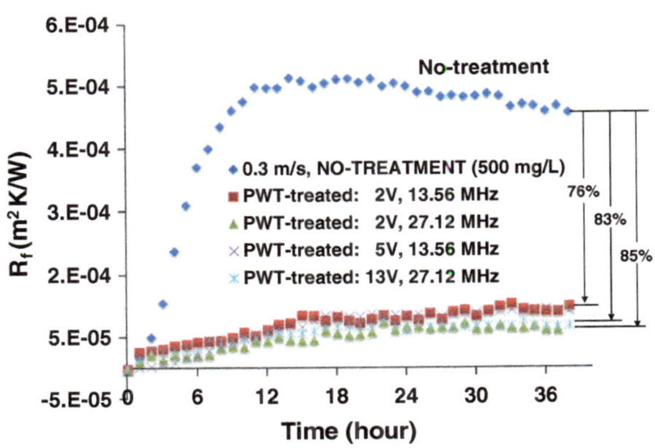

Fig. 3.21 Fouling tests at different voltages and two different frequencies (550 mg/l, 0.3 m/s) [68]

3.5.2.3 Fouling Resistance

Figure 3.21 shows the results for the fouling tests obtained using water hardness of 550 mg/l and a constant velocity of 0.3 m/s for both no-treatment and PWT-treatment cases. The no-treatment case was conducted twice for repeatability check and it was found that both fouling curves looked very consistent with maximum 10 % difference in the asymptotic fouling resistance value at the end of the test. Due to very high water hardness, there was no induction period observed in all tests. An induction period is usually indicated by a straight horizontal line in the beginning of the fouling curve, and is defined as the lateral spreading of scale deposits on heat transfer surface until it is fully covered mainly by stable crystal nucleation [93]. The artificial hard water that contained the artificially-made calcium and bicarbonate ions reacted quickly to the hot heat transfer surface, making immediate deposition on the surface. The scale deposition involved the cumulative effect of direct diffusion of dissolved calcium ions to the heat transfer surface and the deposition of precipitated calcium salt particles due to supersaturated conditions and accelerated precipitation of calcium salts by PWT . The no-treatment case (i.e., without the use of PWT device) had a steep increase in fouling resistance in the first 10 h of operation, indicating that $CaCO_3$ precipitated and continuously adhered to the heat transfer surface. At t = 12 h, the fouling resistance reached an asymptotic level, suggesting that a significant removal of scale particles was taking place, maybe due to reduced opening and subsequently increased shear stress.

The PWT-treatment cases gave very different fouling curves. For example, the slope in the fouling curve during the first 10 h was much smaller than that observed for the no-treatment case. If one assumes that the deposition of new scales for the PWT case was approximately the same as that for the no-treatment case, one could conclude that the removal rate was significantly greater for the PWT case than the no-treatment case. The fouling resistance continued to rise, reaching asymptotic levels at approximately t = 15 h, after which the fouling resistances remained almost unchanged. The PWT-treatment cases at 2 V, 13.56 MHz and 5 V, 13.56 MHz gave similar fouling resistance drops of about 76 % from the no-treatment case value from 4.1×10^{-4} to 9.1×10^{-5} m^2K/W. The best result in the fouling resistance drop was obtained from the case of 2 V and 27.12 MHz where there was an 85 % reduction in the fouling resistance. As the frequency in the oscillating electric field increased, the efficacy of the PWT in the prevention of fouling also increased. One can explain that at a higher frequency, the calcium ions have a greater chance to collide with each other, thus enabling precipitation in the bulk water at a higher rate. The precipitated calcium salt particles in bulk water adhere to the heat transfer surface, but in the soft sludge form of particulate fouling via gravitational settling and particle transport. Particulate fouling is known to be less-adherent which can be relatively easily removed by the shear force produced by flow velocity . Visual inspection on the fouled tubes also confirmed thinner deposits for the PWT-treatment cases as compared with the no-treatment case (Fig. 3.22). It was also noticeable that as the velocity of cold water in the heat transfer test section was increased, the fouling

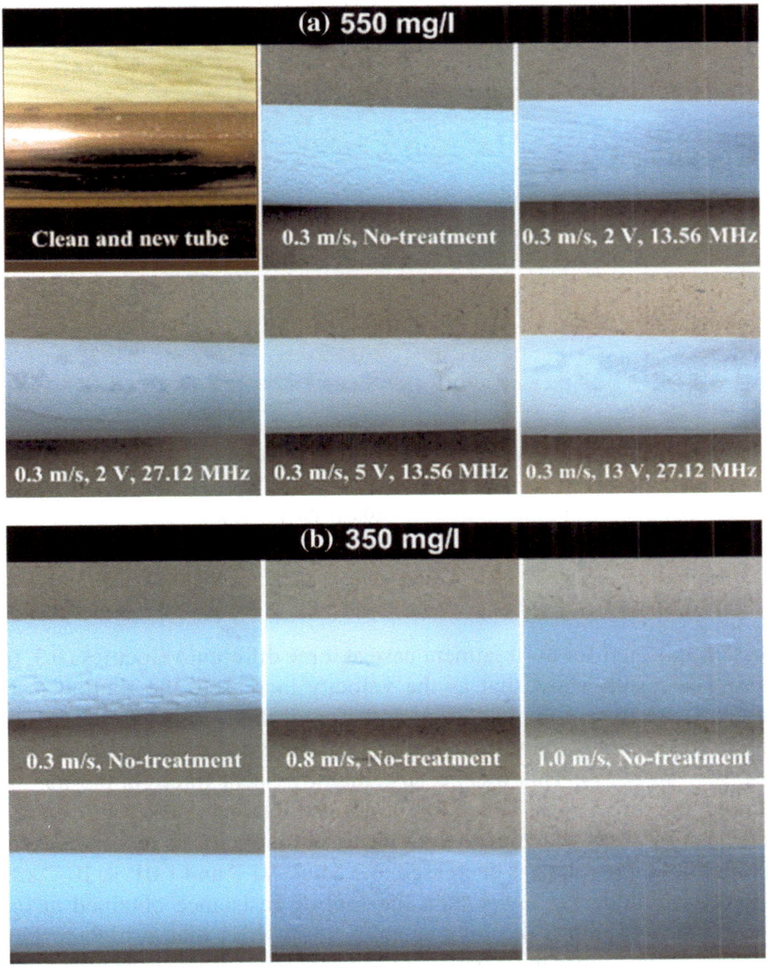

Fig. 3.22 Photographic images of the clean and fouled tubes at artificial $CaCO_3$ hardness [68]

deposit decreased its thickness as expected for both PWT-treatment and no-treatment cases (Fig. 3.22b).

The flow velocity alone can have an effect on the thickness of scale deposition on heat transfer surfaces. Many researchers reported decreased fouling resistance when the flow velocity was increased [99, 100]. Bansal and Muller-Steinhagen [21] reported that the calcium sulphate fouling resistance in plate heat exchangers decreased from 5×10^{-8} to 1×10^{-8} m^{2}K/W when the flow velocity was increased from 25 to 85 cm/s. Knudsen [100] also observed reduced fouling resistance as the velocity was increased in a shell-and-tube heat exchanger. Figure 3.23 presents the effect of flow velocity on the fouling resistance for

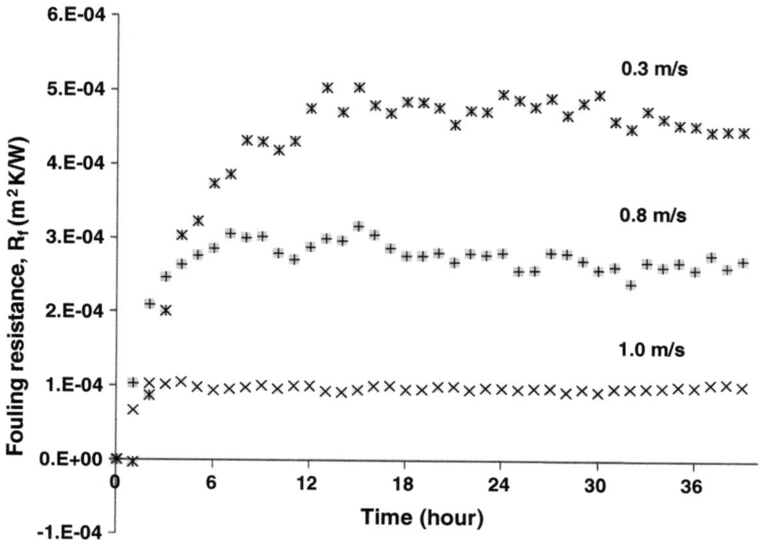

Fig. 3.23 No-treatment case at different flow velocities (350 mg/l $CaCO_3$ hardness) [68]

hardness of 350 mg/l for no-treatment case at three different velocities (0.3, 0.8 and 1.0 m/s). The results show that as the velocity increased, the fouling resistance decreased. There were no induction periods for all three cases. At the lowest flow velocity of 0.3 m/s, the asymptotic fouling resistance was at its highest as expected (4.46×10^{-4} m^2K/W). This was because the flow shear forces at 0.3 m/s was the least among the three cases, which might not be high enough to remove the scales from the heat transfer surface. As velocity was increased to 0.8 m/s and 1.0 m/s, the asymptotic fouling resistances decreased to 2.51×10^{-4} and 1.01×10^{-4} m^2K/W, respectively, significantly lower than the fouling resistance obtained at 0.3 m/s. This supports the findings of previous studies [66, 101] that at higher velocities, even if there was a higher mass deposition rate, the high flow shear forces might be large enough to continuously remove the scales.

The comparison between the no-treatment cases and the PWT-treatment cases at 2 V and 13.56 MHz is shown in Fig. 3.24. The test conditions of 2 V and 13.56 MHz were chosen as they were the best ones among the test conditions used in the present study (refer to Fig. 3.21). The asymptotic fouling resistances dropped by 34–88 % for the PWT cases depending on the velocity of cooling water. The asymptotic fouling resistance at the 0.3 m/s case decreased by about 88 % from the no-treatment case. The high frequency Oscillating electric field is believed to give a greater opportunity for ions to collide with each other compared with the previous PWT methods such as permanent magnets or solenoid coils. Accordingly, the bulk precipitation of calcium particles in water should take place more efficiently than the previous PWT methods. Subsequently, the calcium salt particles suspended in water adhere to the heat transfer surface in the form of particulate fouling.

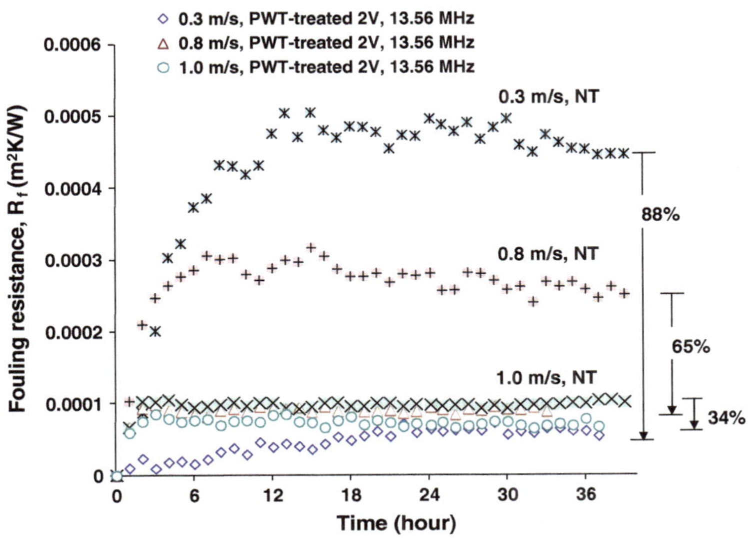

Fig. 3.24 Comparison of the no-treatment and treated cases using two graphite electrode plates at 2 V, 13.56 MHz at different flow velocities (350 mg/l $CaCO_3$ hardness) [68]

Table 3.11 Percentage reduction of fouling resistance by PWT-treatment from the no-treatment case at different velocities (350 mg/l $CaCO_3$ hardness)

Velocity at the cold-water side of the heat transfer test section (m/s)	Equivalent velocity at the PWT device (m/s)	PWT-treatment case R_f reduction from the no-treatment case (%)
0.3	0.12	88
0.8	0.32	65
1.0	0.40	34

As the particulate fouling deposits are known to be soft and not so adherent to the heat transfer surface thus even at low flow velocities (such as 0.3 m/s), the scales could be continuously removed by the shear forces created by the low flow velocity. Table 3.11 presents the percentage of fouling resistance reduction by PWT-treatment case from the no-treatment case at different velocities.

3.5.2.4 SEM Images

The SEM images (5,000×) of the $CaCO_3$ scales obtained for both the no-treatment and PWT-treatment cases are shown in Figs. 3.25 and 3.26. Looking at the SEM images, one can immediately notice the effect of catalytic alloys on the growth of crystals and their microstructures. Figure 3.25 shows the SEM images of the deposits for the case with a water hardness of 550 mg/l and at flow velocity of 0.3 m/s.

Fig. 3.25 SEM photographs (5,000×) of the scales for the **a** no-treatment and **b–e** PWT-treatment cases (550 mg/l $CaCO_3$ hardness and V = 0.3 m/s) [68]

The no-treatment case (Fig. 3.25a) showed sharp and pointed needle-like crystal structures (less than 1 μm in size) while the PWT-treatment cases (Fig. 3.25b–e) showed hexagonal (cubic-like) crystal structures with more than 3 μm in size. The former (no-treatment) depicts the orthorhombic shape of aragonite $CaCO_3$ crystals which are known to be denser and more adherent than calcite (see Table 3.1). The latter (PWT-treatment) are more of calcite crystal structures which are believed to be easily removed by shear forces created by flow. XRD analyses given in Fig. 3.27 confirm the crystal structures. The needle-like structures at the no-treatment case suggests that the crystals might have grown directly on the heat transfer surface through crystallization reaction with additional ions diffusing from the bulk water to the heat transfer surface. On the other hand, for the PWT-treatment cases, the $CaCO_3$ crystals were found to be generally bigger than those for the no-treatment case and appeared to be an accumulation of small-sized crystallites. It can be speculated that submicron-size $CaCO_3$ particles were initially produced in the bulk water by the

Fig. 3.26 SEM photographs (5,000×) of the scales for the no-treatment and PWT-treatment cases at 350 mg/l $CaCO_3$ hardness [68]

oscillating electric field and they eventually grew in size due to continuous circulation in the cooling loop and deposited unto the surface in the form of small particulates, i.e. in the form of particulate fouling. These particulates forming a soft sludge go to the heat transfer surface via gravitational settling or particle transport. Of note is that both no-treatment and PWT-treatment cases were at a constant flow velocity of 0.3 m/s at the heat transfer test section.

Figure 3.26 shows the SEM images at three different velocities for 350-mg/l hard water case. The SEM images for the no-treatment cases with V = 0.3 and V = 0.8 m/s had less than 1 μm in crystal thickness, whereas those obtained with V = 1.0 m/s had less than 3 μm in size. All of the three no-treatment cases showed sharp and pointed crystal structures. The sharp and pointed crystal tips could have increased the attachment capability of the deposit to adhere on the heat

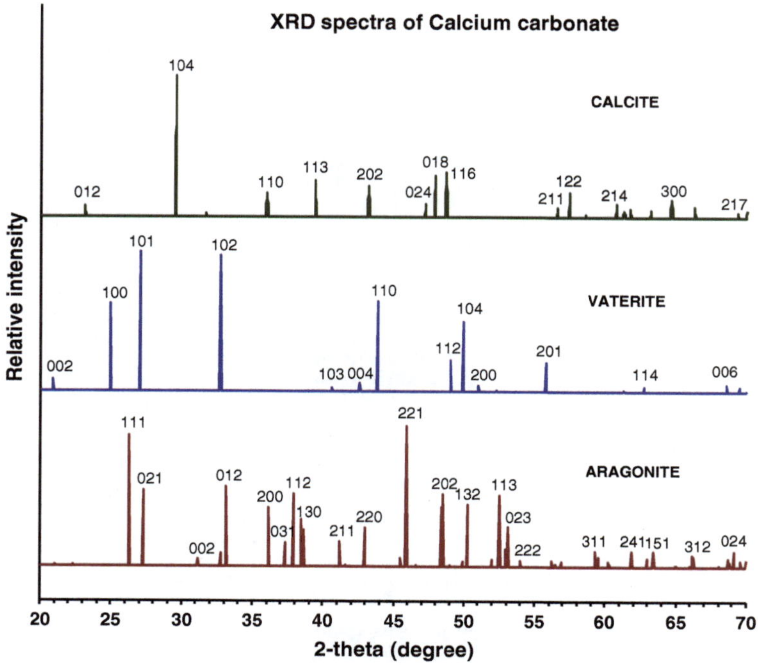

Fig. 3.27 Standard XRD spectra of $CaCO_3$ in three crystal phases: aragonite, vaterite, and calcite [68]

transfer surface more strongly than blunt crystals for the PWT-treatment cases. On the other hand, the PWT-treatment cases (2 V, 13.56 MHz treatment) showed larger crystal structures with blunt and dull crystal surfaces. At V = 0.3 m/s, the crystals looked like a bigger version of the crystals at the no-treatment condition but with dull surface and thickness of more than 4 μm. The crystal structure obtained at V = 0.8 m/s (PWT-treatment) was hexagonal in shape with thickness of over 4 μm. The PWT-treatment crystal structures obtained at V = 1.0 m/s showed flat and flaky-type agglomerated scales. One could see the lesser steepness in the structure of the images of those obtained for the PWT-treatment cases as compared with those for no-treatment cases especially when velocity was increased, presumably due to a larger flow shear forces acting on the deposit. Table 3.12 presents the summary of the test results.

3.5.2.5 X-ray Diffraction

Calcium carbonate is a crystalline substance that exists in three polymorphs: calcite, aragonite and vaterite [101, 102]. Each polymorph has unique crystallographic structure. X-ray diffraction (XRD) is a non-destructive technique to determine the crystallographic structure and chemical composition of a substance

Table 3.12 Summary of the test results for fouling resistance, SEM, and XRD

Parameter	R_f, m^2K/W (Percentage decrease from no-treatment case)	SEM	XRD
550 mg/l			
No-treatment	4.1×10^{-4}	Sharp, pointed, needle-like	Aragonite
2 V, 13.56 MHz	9.1×10^{-5} (↓76 %)	Hexagonal, thick	Calcite
2 V, 27.12 MHz	6.1×10^{-5} (↓85 %)	Hexagonal, thick	Calcite
5 V, 13.56 MHz	9.1×10^{-5} (↓76 %)	Hexagonal, thick	Calcite
13 V, 27.12 MHz	6.9×10^{-5} (↓83 %)	Hexagonal, thick	Calcite
350 mg/l			
No-treatment (0.3 m/s)	4.5×10^{-4}	Sharp, pointed, needle-like	Aragonite
2 V, 13.56 MHz (0.3 m/s)	5.4×10^{-5} (↓88 %)	Hexagonal, thick	Aragonite
No-treatment (0.8 m/s)	2.5×10^{-4}	Sharp, pointed, needle-like	Aragonite
2 V, 13.56 MHz (0.8 m/s)	8.8×10^{-5} (↓65 %)	Hexagonal, thick	Calcite
No-treatment (1.0 m/s)	1.0×10^{-4}	Sharp, pointed, needle-like	Aragonite
2 V, 13.56 MHz (1.0 m/s)	6.7×10^{-5} (↓34 %)	Agglomerated, flat, thick	Aragonite

by recording the intensity of an X-ray as a function of 2θ angle. The crystal structure could be determined by knowing the angles and distances between the atoms of a substance according to Bragg's Law [102]:

$$2d \sin \theta = n\lambda \tag{3.5.1}$$

where d is the distance between atoms, θ is the diffraction angle, n is an integer, and λ is the wavelength of the X-ray. Every substance has unique XRD spectrum that serves as like its fingerprint [102]. Figure 3.27 shows the standard XRD spectra of the three crystal phases of calcium carbonate. Calcite has a prominent peak of intensity at $2\theta = 29.5°$ and Miller index of 104, while aragonite has its peak intensity at $2\theta = 45.9°$ and Miller index of 221. Miller index is a notation to quantify the planes and directions in crystal lattices [102].

Figure 3.28 presents the results of the present XRD analyses for both the no-treatment and PWT-treatment cases at water hardness of 550 mg/l and a flow velocity of 0.3 m/s. The results here were compared to the standard XRD spectra of $CaCO_3$ found in Fig. 3.27. XRD analysis was conducted to determine the crystallographic phase of the deposit so that the focus was on the spectrum peaks and not on the intensity. For the no-treatment case, the peaks depicted that of an aragonite crystal while the results obtained from all the PWT-treatment cases resembled the calcite phase of calcium carbonate. The XRD results here support the SEM image results given in Fig. 3.25. The supersaturated condition in cooling water must have spontaneously produced coagulated $CaCO_3$ particles in bulk water and with the added effect of Oscillating electric fields, more particles must have precipitated in the bulk water either by collision between mineral ions or by attachment of ions to the nucleates. Since they were not directly exposed to the high-temperature heat transfer surface, the precipitated particles could be in the

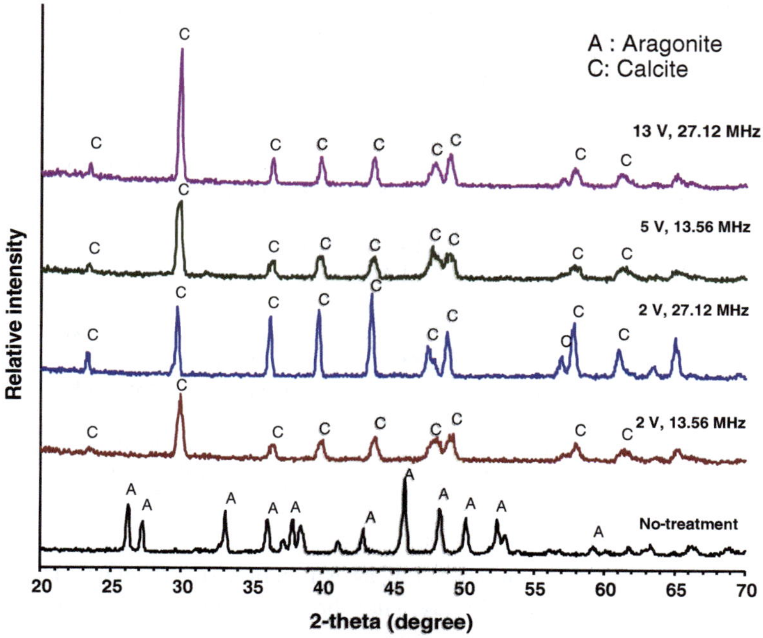

Fig. 3.28 XRD analyses of the scales for the no-treatment and PWT-treatment cases at V = 0.3 m/s and 550 mg/l $CaCO_3$ hardness [68]

form of calcite crystals. On the other hand, the mineral ions that precipitated directly on the hot heat transfer surface are believed generally to form into aragonite crystals, as with the case of no-treatment.

The XRD results for the no-treatment cases at a water hardness of 350 mg/l and at different flow velocities are shown in Fig. 3.29. At the no-treatment case, most of the dissolved ions were believed to directly diffuse and deposit on the heat transfer surface, thus making them more adherent and affinitive on the surface. The scales formed directly on the heat transfer surface (i.e., precipitation fouling) are known to be tenacious and difficult to remove. The no-treatment cases depicted those of aragonite crystal structures, which are known to be a troublesome type of scales. On the other hand, Fig. 3.30 shows the XRD spectra for the PWT-treatment cases at 2 V and 13.56 MHz. The results obtained at 0.3 m/s and 1.0 m/s velocities revealed those of aragonite crystal phase, and the scale was calcite for 0.8 m/s. There are many factors that could affect the morphology of crystals on heat transfer surfaces. Among such were water temperature, hardness, flow velocity at the test section, and the residence time and frequency of oscillating electric fields at the PWT device. The differences in the residence time and frequency of the treatment of cooling water inside the PWT device might have an effect in the coagulation tendency of dissolved mineral ions and particles in the bulk water. At a flow velocity of 0.3 m/s, the residence time of water (thus calcium ions) at the

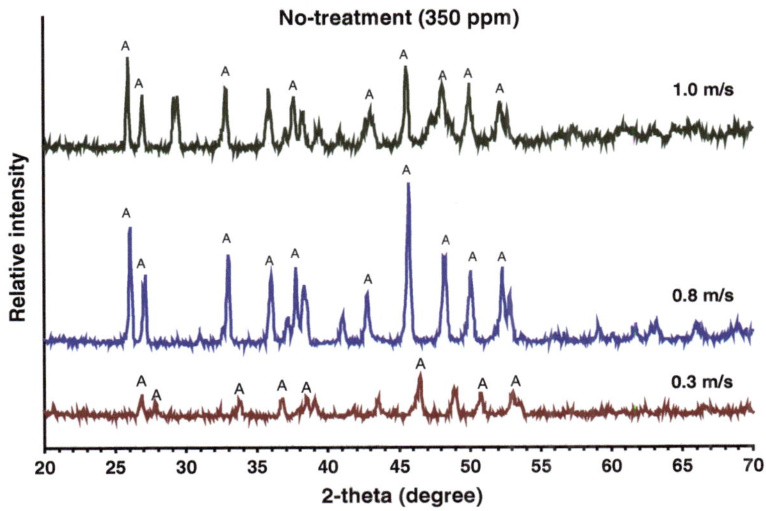

Fig. 3.29 XRD analyses of the scales for the no-treatment cases at varying flow velocities and 350 mg/l $CaCO_3$ hardness [68]

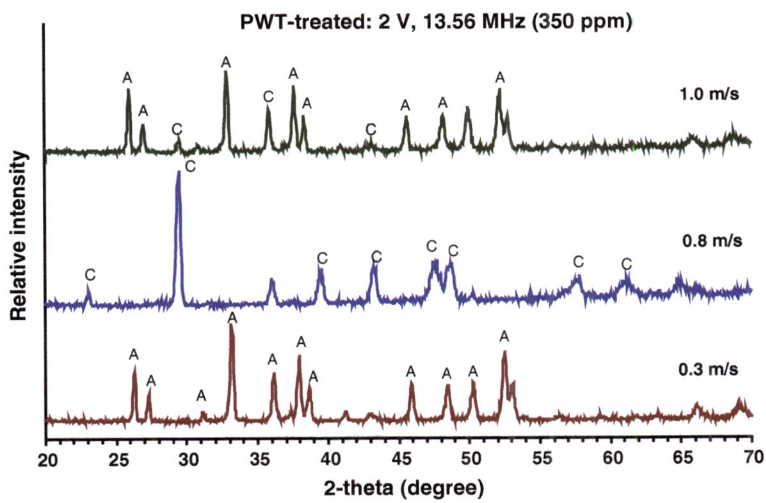

Fig. 3.30 XRD analyses of the scales for the PWT-treatment cases at varying flow velocities and 550 mg/l $CaCO_3$ hardness [68]

PWT device might be long enough, yet the cooling water returned to the PWT device for treatment less frequently compared to greater flow cases. The result at a flow velocity of 0.8 m/s showed calcite form of calcium carbonate. Its residence time and the frequency of treatment at the PWT device might have been just enough for the given hardness of water. It is of note that the hardness in this test was at 350 mg/l which was lower than the 550 mg/l in the first test yet still at significantly supersaturated condition.

The present study investigated the effectiveness of a new physical water treatment method using oscillating electric fields on the mitigation of $CaCO_3$ fouling in a concentric tube heat exchanger. The study utilized oscillating electric fields produced between two graphite electrode plates directly in water. Fouling tests were conducted at two different water hardness (350 and 550 mg/l) using artificial hard water, and at three different cold-water side velocities of 0.3, 0.8 and 1.0 m/s, with the corresponding flow velocities inside the graphite plates of 0.12, 0.32 and 0.40 m/s.

The fouling resistances for the PWT-treatment cases dropped by 34–88 % compared with those obtained for the no-treatment cases, depending on the cold-side flow velocity and the water hardness. SEM images showed hexagonal and agglomerated crystal-shape structures for the PWT-treatment cases as compared to sharp and needle-like structure for the no-treatment cases. The XRD analyses showed aragonite crystals for the no-treatment cases while the PWT-treatment cases showed both aragonite and calcite forms of scale depending on water hardness and flow velocity. The voltage and frequency for the best mitigating performance were found to be 2 V and 13.56 MHz, respectively.

3.6 Summary

Mineral fouling of heat exchangers is a major issue in industrial operation and maintenance. Many methods including chemical and non-chemical means were investigated on their efficacy to mitigate and clean mineral fouling. The non-chemical PWT method is one of the promising methods because of its environment-friendly approach and successful applications.

In this study, a new PWT using high-frequency oscillating electric fields produced directly in water was used to mitigate scaling in heat exchangers. This is a major improvement over the previous PWT methods with low electric field strength (about 1 mV/cm) and low allowable frequency (about 2 kHz). The efficacy of this new PWT to mitigate scaling in a double-pipe heat exchanger was investigated. Artificially-hardened and natural hard water at varying calcium carbonate hardness were used for the fouling tests. Different combinations of voltages and frequencies were investigated to get the optimum values for the mitigation of scaling in heat exchangers. The hypothesis for PWT is bulk precipitation, wherein the high-frequency oscillating electric fields in the present study precipitate the dissolved mineral ions such as calcium to mineral salts in bulk water. This bulk precipitation due to PWT leads to the formation of soft sludge or particulate fouling. This type of fouling is believed to easily be removed by shear forces created by flow than those deposits produced from the precipitation of mineral ions directly on the solid heat transfer surface, i.e., precipitation fouling.

Bulk precipitation tests showed that PWT-treatment increased the number of particles up to five times than the no-treatment case depending on the flow rate. When the water is exposed to PWT more often, one can have higher precipitation of particles. The fouling resistance of PWT-treatment had a maximum decrease of 80 % from the no-treatment case, as evidenced by the thinner scale deposits after the fouling tests. The fouling deposits exhibited the minimum thickness at high velocity. Since PWT works via bulk precipitation and particulate fouling mechanism, a high velocity flow, i.e. a high shear force, is needed so that the deposits could effectively be removed. This is to say that PWT will not work without enough flow shear forces. Furthermore, PWT is most effective for re-circulating water, because the higher the exposure of the circulating water to PWT, the better is its effectiveness. The new PWT method using high-frequency oscillating electric fields presents a valid tool to mitigate scaling in heat exchangers for cooling-water applications.

Acknowledgments The authors would like to acknowledge the support of the National Research Foundation of Korea, the faculty of the Division of Mechanical Design Engineering, and friends and colleagues at the Physical Water Treatment and Biosystems Laboratory at Chonbuk National University, Korea.

References

1. T.R. Bott, *Fouling of Heat Exchangers* (Elsevier, The Netherlands, 1995)
2. E.F.C. Somerscales, Fouling of heat transfer surface: an historical review. Heat Transf. Eng. **11**(1), 19–36 (1990)
3. W. Li, R.L. Webb, Fouling in enhanced tubes using cooling tower water Part II: combined particulate and precipitation fouling. Int. J. Heat Mass Transf. **43**(19), 3579–3588 (2000)
4. T.Q. Liu, X.Q. Li, H.L. Wang, X.Y. Sun, Formation process of mixed fouling of microbe and $CaCO_3$ in water systems. Chem. Eng. J. **88**(1–3), 249–254 (2002)
5. Y.I. Cho, S.H. Lee, W. Kim, Physical water treatment for the mitigation of mineral fouling in cooling-water applications. ASHRAE **109**, 346–357 (2003)
6. Y.I. Cho, A.F. Fridman, S.H. Lee, W.T. Kim, Physical water treatment for fouling prevention in heat exchangers. Adv. Heat Transf. **38**, 1–71 (2004)
7. C.B. Panchal, K. JG, Mitigation of water fouling: technology status and challenges. Adv. Heat Transf. **31**, 431–474 (1998)
8. T.R. Bott, Aspects of crystallization fouling. Exp. Therm. Fluid Sci. **14**(4), 356–360 (1997)
9. T. Kuppan, *Heat Exchanger Design Handbook* (Marcel Dekker, New York, 2000)
10. The U.S. Department of Energy, Non-chemical technologies for scale and hardness control. DOE/EE-0162. (1998)
11. R. Steinhagen, H. Mullersteinhagen, K. Maani, Problems and costs due to heat-exchanger fouling in New-Zealand industries. Heat Transf. Eng. **14**(1), 19–30 (1993)
12. S. Radler, U. Ousko-Oberhoffer, Optimised heat exchanger management—achieving financial and environmental targets. in *Proceedings of the 6th International Conference on Heat Exchanger Fouling and Cleaning—Challenges and Opportunities RP2*, 2005
13. H. Muller-Steinhagen, *Handbook of Heat Exchanger Fouling—Mitigation and Cleaning Technologies* (Publico Publications, Germany, 2000)
14. L.D. Tijing, H.Y. Kim, D.H. Lee, C.S. Kim, Y.I. Cho, Use of an oscillating electric field to mitigate mineral fouling in a heat exchanger. Exp. Heat Transf. **22**(4), 257–270 (2009)

15. Y.I. Cho, J. Lane, W. Kim, Pulsed-power treatment for physical water treatment. Int. Commun. Heat Mass **32**(7), 861–871 (2005)
16. K.D. Demadis, E. Mavredaki, A. Stathoulopoulou, E. Neofotistou, C. Mantzaridis, Industrial water systems: problems, challenges and solutions for the process industries. Desalination **213**(1–3), 38–46 (2007)
17. J. Lane, Y.I. Cho, W. Kim, Pulsed-power water treatment as a green scale inhibitor for HVAC and once-through industrial systems. Corrosion Paper **04541**, 1–21 (2004)
18. L.D. Tijing, B.C. Pak, D.H. Lee, Y.I. Cho, Heat-treated titanium balls for the mitigation of mineral fouling in heat exchangers. Exp. Heat Transf. **21**(2), 115–132 (2008)
19. K.J. Kronenberg, Physical water treatment de-mystified. Magnets 6–15 (1986)
20. S.S. Al-Jaroudi, A. Ul-Hamid, J.A. Al-Matar, Prevention of failure in a distillation unit exhibiting extensive scale formation. Desalination **260**(1–3), 119–128 (2010)
21. B. Bansal, H. Muller-Steinhagen, Crystallisation fouling in plate heat exchangers. ASME J. Heat Transf. **115**, 584–591 (1993)
22. A. Helalizadeh, H. Muller-Steinhagen, M. Jamialahmadi, Mixed salt crystallisation fouling. Chem. Eng. Process. **39**(1), 29–43 (2000)
23. M. G. Mwaba, Analysis of heat exchanger fouling in cane industry. Dissertation, Eindhoven University of Technology, The Netherlands, 2003
24. D. Hasson, Rate of decrease of heat transfer due to scale deposition. Dechema-Monographien **47**, 233–252 (1962)
25. F.P. Incropera, D.P. DeWitt, *Fundamentals of Heat and Mass Transfer*, 5th edn. (Wiley, New York, 2002)
26. J. Nesta, C.A. Bennet, Fouling mitigation by design. in *Proceedings of the 6th International Conference on Heat Exchanger Fouling and Cleaning—Challenges and Opportunities RP2*, 5–10 June 2005
27. D.J. Kukulka, M. Devgun, Fluid temperature and velocity effect on fouling. Appl. Therm. Eng. **27**(16), 2732–2744 (2007)
28. W.T. Kim, C. Bai, Y.I. Cho, A study of $CaCO_3$ fouling with a microscopic imaging technique. Int. J. Heat Mass Transf. **45**(3), 597–607 (2002)
29. H. Muller-Steinhagen, Cooling water fouling in heat exchangers. Adv. Heat Transf. **33**, 415–496 (1999)
30. W.T. Kim, Y.I. Cho, Experimental study of the crystal growth behavior of $CaCO_3$ fouling using a microscope. Exp. Heat Transf. **13**(2), 153–161 (2000)
31. A.P. Watkinson, O. Martinez, Scaling of heat exchanger tubes by calcium carbonate. Trans. ASME **97**, 504–508 (1975)
32. D. Hasson, M. Avriel, W. Resnick, T. Rozenman, S. Windreich, Mechanism of calcium carbonate scale deposition on heat-transfer surface. Ind. Eng. Chem. Fund **7**, 58–63 (1968)
33. J.C. Cowan, D.J. Weintritt, *Water-Formed Scale Deposits* (Gulf-Publishing Company, Houston, 1976)
34. J. MacAdams, S. Parsons, Calcium carbonate scale formation and control. Rev. Environ. Sci. Bio/Technol. **3**, 159–169 (2004)
35. D. Chapman, *A Guide to the Use of Biota, Sediments and Water in Environmental Monitoring* (Chapman Hall, London, 1992)
36. V. Snoeyink, D. Jenkins, *Water Chemistry* (Wiley, New York, 1980)
37. R. Ketrane, B. Saidani, O. Gil, L. Leleyter, F. Baraud, Efficiency of five scale inhibitors on calcium carbonate precipitation from hard water: effect of temperature and concentration. Desalination **249**(3), 1397–1404 (2009)
38. T. Casanueva-Robles, T. Bott, The environmental effect of heat exchanger fouling: a case study. in *Proceedings of the 6th International Conference on Heat Exchanger Fouling and Cleaning—Challenges and Opportunities RP2*, 5–10 June 2005
39. L. Melo, T. Bott, C. Bernardo, Fouling science and technology E: applied science. in *NATO ASI Series 145* (Kluwer, The Netherlands, 1988)

40. D.M. Yebra, S. Kiil, K. Dam-Johansen, Antifouling technology—past, present and future steps towards efficient and environmentally friendly antifouling coatings. Prog. Org. Coat. **50**(2), 75–104 (2004)
41. S.A. Parsons, S.J. Judd, T. Stephenson, S. Udol, B.L. Wang, Magnetically augmented water treatment. Process Saf. Environ. **75**(B2), 98–104 (1997)
42. J. Donaldson, S. Grimes, Lifting the scales from our pipes. New Sci. **117**(1600), 43–46 (1988)
43. C. Smith, P.P. Coetzee, J.P. Meyer, The effectiveness of a magnetic physical water treatment device on scaling in domestic hot-water storage tanks. Water SA **29**(3), 231–236 (2003)
44. J.S. Baker, S.J. Judd, Magnetic amelioration of scale formation. Water Res. **30**(2), 247–260 (1996)
45. A. Fathi, T. Mohamed, G. Claude, G. Maurin, B.A. Mohamed, Effect of a magnetic water treatment on homogeneous and heterogeneous precipitation of calcium carbonate. Water Res. **40**(10), 1941–1950 (2006)
46. X.K. Xing, Research on the electromagnetic anti-fouling technology for heat transfer enhancement. Appl. Therm. Eng. **28**(8–9), 889–894 (2008)
47. X.K. Xing, C.F. Ma, Y.C. Chen, Investigation on the electromagnetic anti-fouling technology for scale prevention. Chem. Eng. Technol. **28**(12), 1540–1545 (2005)
48. Z.H. Quan, Y.C. Chen, C.F. Ma, C.M. Wang, B. Li, Experimental study on anti-fouling performance in a heat exchanger with low voltage electrolysis treatment. Heat Transf. Eng. **30**(3), 181–188 (2009)
49. G. Ushakov, Antiscaling treatment of water by an electric field in heat-supply networks. Therm. Eng. **55**(7), 570–573 (2008)
50. A. Shahryari, M. Pakshir, Influence of a modulated electromagnetic field on fouling in a double-pipe heat exchanger. J. Mater. Process Tech. **203**(1–3), 389–395 (2008)
51. Y.I. Cho, B.G. Choi, B.J. Drazner, Electronic anti-fouling technology to mitigate precipitation fouling in plate-and-frame heat exchangers. Int. J. Heat Mass Transf. **41**(17), 2565–2571 (1998)
52. G.J. Lee, L.D. Tijing, B.C. Pak, B.J. Baek, Y.I. Cho, Use of catalytic materials for the mitigation of mineral fouling. Int. Commun. Heat Mass **33**(1), 14–23 (2006)
53. P.P. Coetzee, M. Yacoby, S. Howell, S. Mubenga, Scale reduction and scale modification effects induced by Zn and other metal species in physical water treatment. Water SA **24**(1), 77–84 (1998)
54. J. Macadam, S.A. Parsons, Calcium carbonate scale control, effect of material and inhibitors. Water Sci. Technol. **49**(2), 153–159 (2004)
55. Q.G. Tang, J.P. Meng, J.S. Liang, L. Nie, Y.X. Li, Effects of copper based alloys on the nucleation and growth of calcium carbonate scale. J. Alloy. Compd. **491**(1–2), 242–247 (2010)
56. L.D. Tijing, M.H. Yu, C.H. Kim, A. Amarjargal, Y.C. Lee, D.H. Lee, D.W. Kim, C.S. Kim, Mitigation of scaling in heat exchangers by physical water treatment using zinc and tourmaline. Appl. Therm. Eng. **31**(11–12), 2025–2031 (2011)
57. H. Al-Qahtani, Effect of magnetic treatment on gulf seawater. Desalination **107**, 75–81 (1996)
58. S. Kazi, G. Duffy, X. Chen, Mineral scale formation and mitigation on metals and a polymeric heat exchanger surface. Appl. Therm. Eng. **30**(14–15), 2236–2242 (2010)
59. L. Tijing, B. Pak, B. Baek, D. Lee, Y. Cho, An experimental study on the bulk precipitation mechanism of physical water treatment for the mitigation of mineral fouling. Int. Commun. Heat Mass Trans. **34**, 673–681 (2007)
60. N. Kim, R. Webb, Particulate fouling of water in tubes having a two-dimensional roughness geometry. Int. J. Heat. Mass. Trans. **34**, 2727–2738 (1991)
61. E. Somerscales, A. Ponteduro, A. Bergles, Particulate of heat transfer tubes enhanced on their inner surface. ASME Fouling Enhan. Inter. **164**, 17–28 (1991)

62. O. Sohnel, J. Mullin, Some comments on the influence of a magnetic field on crystalline scale formation. Chem. Ind. **6**, 356–358 (1988)
63. D. Lisitsin, Q. Yang, D. Hasson, R. Semiat, Inhibition of $CaCO_3$ scaling on RO membranes by trace amounts of zinc ions. Desalination **183**, 289–300 (2005)
64. C. Gabrielli, R. Jaouhari, G. Maurin, M. Keddam, Magnetic water treatment for scale prevention. Wat Res. **35**(13), 3249–3259 (2001)
65. F. Alimi, M. Tlili, M. Ben Amor, C. Gabrielli, G. Maurin, Influence of magnetic field on calcium carbonate precipitation. Desalination **206**, 163–168 (2007)
66. J. Kim, J. Jung, I. Yeom, G. Aoh, Electric fields treatment for the reduction of membrane fouling, the inactivation of bacteria and the enhancement of particle coagulation. Desalination **202**, 31–37 (2007)
67. S. Lee, A study of physical water treatment technology to mitigate the mineral fouling in a heat exchanger. Dissertation, Drexel University, Philadelphia, PA, 2002
68. L.D. Tijing, H.Y. Kim, D.H. Lee, C.S. Kim, Y.I. Cho, Physical water treatment using RF electric fields for the mitigation of $CaCO_3$ fouling in cooling water. Int. J. Heat Mass Tran. **53**(7–8), 1426–1437 (2010)
69. K. Busch, M. Busch, Laboratory studies on magnetic water treatment and their relationship to a possible mechanism for scale reduction. Desalination **109**, 131–148 (1997)
70. D. Parker, An investigation of the role of magnetic water treatment devices in calcium carbonate scale formation. MS thesis, Baylor University, Texas, 1985
71. J. Grutsch, J. McClintock, Corrosion and deposit control in alkaline cooling water using magnetic water treatment at AMOCO's largest refinery. Corrosion **84**(330), 1–26 (1984)
72. J. Bogatin, Magnetic treatment of irrigation water: experimental results and application conditions. Environ. Sci. Technol. **33**, 1280–1285 (1999)
73. K. Kronenberg, Experimental evidence for effects of magnetic fields on moving water. IEEE Trans. Magnet **MAG-21**, 2059–2061 (1985)
74. L. Lipus, J. Krope, L. Garbai, Magnetic water treatment for scale prevention. Hungarian J. Indus. Chem. **22**, 239–242 (1994)
75. R. Marth, A scientific definition of the magnetic treatment of water: Its subsequent use in preventing scale formation and removing scale. MTW Introduction 6.1 (1997)
76. R. Szostak Magnetic fluid conditioning system allows 3000 ppm hardness without cooling tower scale buildup. Chem. Proc. 44–45 (1985)
77. R. Diamant, Magnetic treatment of water. Hosp. Eng. **24**, 231 (1970)
78. T. Vermeiren, Magnetic treatment of liquids for scale and corrosion prevention. Cor. Technol. **5**, 215–219 (1958)
79. V. Belova, Magnetic treatment of water. Sov. Sci. Rev. **90**, 150–156 (1972)
80. Y. Wang, J. Babchin, L. Chernyi, R. Chow, R. Sawatzky, Rapid onset of calcium carbonate crystallization under the influence of a magnetic field. Wat. Res. **31**, 346–350 (1997)
81. S. Nasrazadani, T. Chao Laboratory evaluations of ozone as a scale inhibitor for use in open recirculating cooling systems. ASHRAE Research Project 765-RP, Final Report (1994)
82. R. Sheikholeslami, A. Watkinson, Scaling of plain and externally finned heat exchanger tubes. J. Heat Trans. **108**, 147–152 (1986)
83. R. Morse, J. Knudsen, Effect of alkalinity on the scaling of simulated cooling tower water. Can. J. Chem. Eng. **55**, 272–278 (1977)
84. T. Paakkonen, M. Riihimaki, E. Puhukka, E. Muurinen, C. Simonson, R. Keiski, Crystallization fouling of $CaCO_3$—Effect of bulk precipitation on mass deposition on the heat transfer surface. in *Proceedings of International Conference on Heat Exchanger Fouling and Cleaning VII, Schladming*, (Austria 2009)
85. Federal Communications Commission, Operating frequencies. Code of Federal Regulations 1(18):301 (2007)
86. Y. Cho, W. Kim, D. Cho, Electro-flocculation mechanism of physical water treatment for the mitigation of mineral fouling in a heat exchanger. Exp. Heat Trans. **20**(4), 323–335 (2007)

87. R. Gehr, Z.A. Zhai, J.A. Finch, S.R. Rao, Reduction of soluble mineral concentrations in $CaSO_4$ saturated water using a magnetic-field. Water Res. **29**(3), 933–940 (1995)
88. L.D. Tijing, D.H. Lee, D.W. Kim, Y.I. Cho, C.S. Kim, Effect of high-frequency electric fields on calcium carbonate scaling. Desalination **279**(1–3), 47–53 (2011)
89. G. Hearn, *Static Electricity: Guidance for Plant Engineers. Graham Hearn–Wolfson Electrostatics* (University of Southampton, 2002)
90. W. Li, Modeling liquid-side particulate fouling in internal helical-rib tubes. Chem. Eng. Sci. **62**, 4204–4213 (2007)
91. F. Kline, F. McClintock, Describing uncertainties in single sample experiments. Mech. Eng. **75**, 3–8 (1953)
92. C. Fan, Y. Cho, Microscopic observation of calcium carbonate particles: validation of an electronic anti-fouling technology. Int. Commun. Heat Mass Trans. **24**(6), 747–756 (1997)
93. Q. Yang, Y. Liu, A. Gu, J. Ding, Z. Shen, Investigation of induction period and morphology of $CaCO_3$ fouling on heated surface. Chem. Eng. Sci. **57**, 921 (2002)
94. S. Parsons, B. Wang, S. Judd, T. Stephenson, Magnetic treatment of calcium carbonate scale—effect of pH control. Wat. Res. **31**(2), 339–342 (1997)
95. N. Andritsos, A. Karabelas, The influence of particulates on $CaCO_3$ scale formation. J. Heat Trans. **121**, 225–227 (1999)
96. D. Dawson, A non-chemical water treatment system. Corros. Prev. Contr. 61–64 (1990)
97. B. Hauser, *Drinking Water Chemistry: A Laboratory Manual* (CRC Press, LLC, Florida, 2002)
98. E. Chibowski, L. Hotysz, A. Szczes, Time dependent changes in zeta potential of freshly precipitated calcium carbonate. Colloid Surf. A **222**(1–3), 41–54 (2003)
99. S. Lee, Y. Cho, Velocity effect on electronic-antifouling technology to mitigate mineral fouling in enhanced-tube heat exchanger. Int. J. Heat Mass Trans. **45**, 4136–4174 (2002)
100. J. Knudsen, Conquer cooling-water fouling. Chem. Eng. Prog. **87**(4), 42–48 (1991)
101. C. Kontoyannis, N. Vagena, Calcium carbonate phase analysis using XRD and FT-Raman spectroscopy. Analyst **125**, 251–255 (1999)
102. W. Smith, *Principles of Materials Science and Engineering* (McGraw-Hill, Singapore, 1986)

Chapter 4
Flow Patterns, Void Fraction and Pressure Drop in Gas-Liquid Two Phase Flow at Different Pipe Orientations

Afshin J. Ghajar and Swanand M. Bhagwat

Abstract This chapter presents an insightful discussion on flow patterns, void fraction and phenomenon of two phase frictional pressure drop in gas-liquid two phase flow. The flow structure of different flow patterns observed in gas-liquid two phase flow at various pipe orientations are described with the aid of flow visualization. This chapter is helpful in understanding the impact of varying flow patterns, pipe diameter and pipe orientation on the void fraction and two phase pressure drop. Additionally, a brief overview of the void fraction, its dependency on the flow patterns and its influence on the hydrostatic pressure drop is presented. A brief synopsis of the two phase void fraction and frictional pressure drop correlations available in the literature is presented. The performance of these correlations is assessed against a comprehensive database for air-water and refrigerant two phase flow conditions. Based on this detailed analysis, the top performing void fraction and pressure drop correlations are identified and recommended for use for these fluid combinations in different two phase flow situations. Finally, application of the recommended correlations is presented in the form of solved problems. It is expected that these solved problems will give readers an idea of selection and implementation of appropriate correlations for different two phase flow conditions.

Keywords Flow patterns · Void fraction · Drift flux model · Hydrostatic pressure drop · Frictional pressure drop · Gas-liquid two phase flow · Two phase frictional multiplier · Two phase dynamic viscosity

Nomenclature

B_{Ch}	Variable in (Chisholm [20]) correlation
C_o	Distribution parameter
D	Pipe diameter (m)
f	Friction factor

A. J. Ghajar (✉) · S. M. Bhagwat
School of Mechanical and Aerospace Engineering, Oklahoma State University, Stillwater, OK 74078, USA
e-mail: afshin.ghajar@okstate.edu

L. Cheng (ed.), *Frontiers and Progress in Multiphase Flow I*,
Frontiers and Progress in Multiphase Flow, DOI: 10.1007/978-3-319-04358-6_4,

Fr	Froude number ($Fr = G^2/(gD\rho^2)$)
g	Acceleration due to gravity (9.81 m/s^2)
G	Mass flux (kg/m^2 s)
Ku	Kutateladze number as defined by Takeuchi et al. [97]
L	Pipe length (m)
La	Laplace number ($La = \sqrt{\sigma/g(\rho_l - \rho_g)}/D$)
$N_{\mu f}$	Viscosity number $\left(N_{\mu f} = \mu_l/\left(\rho_l\sigma\sqrt{\sigma/g\Delta\rho}\right)^{0.5}\right)$
P	Pressure (Pa)
Re	Reynolds number ($Re = (GD)/\mu$)
S	Slip ratio
U	Phase velocity (m/s)
U_{gm}	Drift velocity (m/s)
We	Weber number ($We = (G^2D)/(\sigma\rho)$)
x	Two phase quality
X	(Lockhart and Martinelli [66]) parameter

Greek Symbols

α	Void fraction
β	Gas volumetric flow fraction
ρ	Phase density (kg/m^3)
μ	Phase dynamic viscosity (Pa-s)
θ	Pipe orientation (degrees)
σ	Surface tension (N/m)
Φ^2	Two phase frictional multiplier

Subscripts

a	Accelerational
atm	Atmospheric
crit	Critical
f	Frictional
g	Gas
go	Gas only
h	Hydrostatic
in	Inlet
j	Phase
l	Liquid
lo	Liquid only
m	Mixture
out	Outlet
s	Superficial
sys	System
t	Total

tp Two phase
tt Turbulent-turbulent
w Water

Superscript
* Non-dimensional parameter

4.1 Introduction

Gas-liquid two phase flow finds its extensive application in the industrial processes pertaining to oil-gas, chemical, nuclear and refrigeration industries. The two component two phase flow referred to as non-boiling flow or two phase flow without undergoing phase change is often encountered in industrial applications such as artificial lift systems and simultaneous transportation of oil and natural gas from remote extraction locations to the processing units. Chemical operations requiring flow of two chemical species together for enhanced mass transfer as in case of ozone treatment of water also relies on gas-liquid non-boiling two phase flow phenomenon. One component two phase flow i.e., the boiling/condensation flow or two phase flow undergoing a phase change process is always present in the nuclear safety operations required for cooling of nuclear reactor rods and in heat exchangers where refrigerants undergo phase change process in evaporators and condensers. For refrigerant two phase flow, the correct estimation of void fraction is required to determine the refrigerant charge inventory while the correct estimation of pressure drop is desired as it influences the refrigerant saturation temperature and hence the overall system efficiency. Irrespective of the type of flow i.e., boiling and non-boiling, for a design engineer the proper understanding of two phase flow phenomenon is of utmost importance for designing of these operations and sizing of process equipment.

For any industrial process from design point of view, the total pressure drop that occurs in the system is of concern which in turn depends strongly on correct understanding of the flow patterns and accurate prediction of the void fraction. The first step in prediction of total pressure drop is the estimation of hydrostatic two phase pressure drop that requires understanding of the flow pattern and accurate estimation of the void fraction at any given pipe orientation, pipe diameter and flow rates of individual phases. The knowledge of flow patterns is necessary since the frictional pressure drop in two phase flow is sensitive to the distribution of gas and liquid phase across the pipe cross section and along the pipe length. Both the flow patterns and void fraction are observed to change with change in flow patterns and pipe orientation and thus result in significantly different pressure drop even measured at similar phase mass flow rates but at different pipe orientations. In order to address these issues with better understandings, one of the objectives of this work is to discuss the effect of different two phase flow variables on the flow patterns, void fraction and hence the pressure drop in two phase flow.

So far an extensive research has been carried out in the general area of two phase flow that has contributed in improving overall understanding of two phase flow phenomenon, development of theories and models to predict flow patterns, void fraction and pressure drop. However, all these developments are confined to certain two phase flow scenarios in terms of pipe geometry and orientations, fluid physical properties and flow patterns and hence fail to predict the desired outcome correctly if arbitrarily used against different two phase flow conditions. Moreover, the majority of this research is focused on the gas-liquid two phase flow in horizontal and vertical upward pipe orientation while comparatively rare information is available on the behavior of two phase flow in upward and downward inclined pipes. The challenging part of solving a two phase flow problem is correct identification of the flow pattern for given set of flow conditions since, the flow patterns are defined on a qualitative basis without any physical parameter to quantify the occurrence of a particular flow pattern at given phase flow rates, pipe geometry and fluid physical properties. The definition of flow pattern depends upon an individual's perception and hence involves the risk of using a wrong correlation to predict the void fraction and pressure drop due to incorrect recognition of the flow pattern.

Thus it is strongly desired to have correlations that can predict void fraction and two phase frictional pressure drop independent of the flow pattern, pipe orientation and pipe diameter. To explore and discuss this issue, the scope of this chapter includes the performance assessment of different correlations that can predict void fraction and two phase frictional pressure drop independent of the flow patterns. The performance of these correlations is assessed against a comprehensive data bank for two phase void fraction and pressure drop in different pipe diameters and pipe orientations available with Two Phase Flow Lab, Oklahoma State University. Based on this assessment, the top three performing correlations are identified for various two phase flow conditions and are recommended to use in prediction of void fraction and two phase frictional pressure drop.

4.2 Flow Patterns and Flow Patterns Map

The flow patterns in gas-liquid two phase flow are generated due to significantly different physical properties of gas and liquid phase, compressibility nature of the gas phase and their alignment with respect to each other across the pipe cross section. The knowledge of flow patterns and flow pattern maps proves instrumental in the general understanding of the physical structure and the mechanism of momentum and heat transfer in two phase flow. The flow pattern map serves as a tool to estimate the sequence of the appearance of different flow patterns with change in the gas and liquid flow rates for a given set of flow conditions. The definitions of flow patterns and their transitions are highly qualitative in nature and are mostly based on the individual's perception. Although some researchers have attempted to develop quantitative methods such as the probabilistic flow regime mapping to predict the existence of certain flow patterns these methods are not universal and are mostly limited to certain flow conditions. The purpose of this section is to present an overview of the flow patterns and the similarities and differences in the physical structure of flow patterns observed in horizontal, vertical upward and vertical downward pipe orientations. The flow pattern maps presented in this section are exclusively developed for two phase flow of air water through a 12.5 mm I.D. pipe at different orientations and are not generic. The objective of introducing these flow patterns maps is to give readers an idea about the effect of pipe orientation on the transition between different flow patterns for fixed pipe diameter and fluid combination.

Let's first look at the physical structure of two phase flow patterns at different pipe orientations. The vertical upward and downward flow share some flow patterns such as bubbly, slug, froth and annular flow in common while the churn flow is observed in vertical upward flow and falling film is observed only in vertical downward orientation. The horizontal flow direction allows existence of bubbly, slug, stratified, wavy and annular flow patterns. Bhagwat and Ghajar [13] illustrated that the two extreme opposite flow directions namely, vertical up and vertical down pipe orientations significantly influence the general appearance of the flow patterns and their transition from one to another. The flow visualization of the bubbly and slug flow show a significant influence of the flow direction on the alignment of one phase with respect to another. These differences in the physical structure of the flow patterns are essentially due to the interaction between buoyancy, gravity and inertia forces. As shown in Fig. 4.1, in case of the vertical upward bubbly flow, the bubbles are distributed evenly across the pipe cross section in the continuous liquid medium for almost all combinations of liquid and gas flow rates. Whereas, the bubbles in vertical downward flow tend to flow in a region close to the pipe center due to repulsion exerted from the pipe wall (known as coring phenomenon) as a function of interaction between buoyancy and inertia forces. With increase in the liquid and gas flow rates, the inertia forces supersede the buoyancy effects and hence the wall repulsion force. Consequently, bubbles start to move towards pipe wall and get more evenly distributed throughout the pipe cross section and finally the physical appearance of the bubbly flow is similar in both vertical upward and downward flow directions. Nguyen [76], Oshinowo [80] and Usui and Sato [102] reported similar type of observations for the coring phenomenon in 19, 25 and 45 mm I.D. vertical downward pipes. It should be noted that the coring phenomenon occurs only in vertical downward pipe orientation and is prominent for large diameter pipes. It is anticipated that the coring phenomenon vanishes with decrease in the pipe diameter since two phase flow literature doesn't provide any evidence of its existence for pipe diameters typically $D < 12$ mm. This difference in general appearance of the bubbly flow is observed to have noticeable effect on the two phase frictional pressure drop in vertical upward and downward flows. In case of the horizontal two phase flow, the bubbles are always observed to flow in the vicinity of the pipe upper wall due to the buoyancy effects

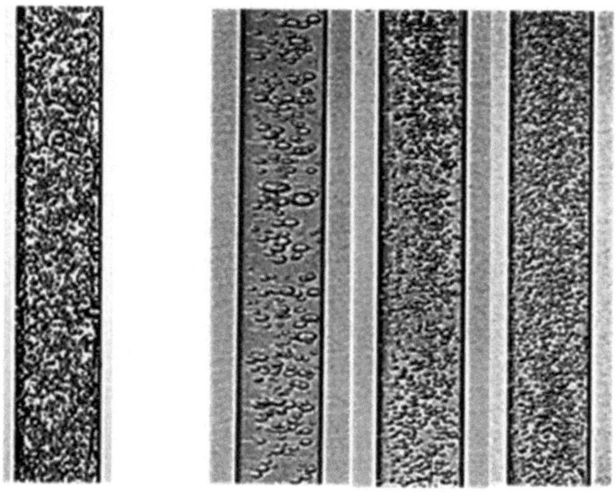

Fig. 4.1 Physical appearance of bubbly flow in vertical upward and downward pipe orientation

Fig. 4.2 Two phase bubbly flow in horizontal pipe orientation

as shown in Fig. 4.2. At fixed gas flow rate, increase in the liquid flow rate results in the shearing of gas bubbles that reduces its size (Figs. 4.2a, b), and consequently increase the number of bubbles that try to penetrate the single phase liquid and enter the near pipe axis region (Fig. 4.2c).

The slug flow regime in vertical upward, downward and horizontal pipe orientations resembles in general appearance i.e., the slug flow in all of these three orientations is characterized by the alternate flow of liquid and elongated gas bubbles. However, a close observation reveals some differences in the shape of the nose and tail region of the slug in addition to the effect of pipe orientation on slug flow direction. The slug in vertical upward orientation appears as a bullet shaped elongated gas bubble pointing in upward direction (due to buoyancy acting in flow

Fig. 4.3 Slug flow in vertical downward two phase flow

direction) while the slug in vertical downward flow can have three possible shapes. Bhagwat and Ghajar [13] have identified that the slug flow in vertical downward orientation initially shows a blunt bubble nose pointing in upward direction for the low values of gas and liquid flow rates. As shown in Fig. 4.3, with increase in both gas and liquid flow rates, the bubble nose becomes flat (Figs. 4.3a, b) and finally appears pointing downwards in the direction of net flow with further increase in the flow rates (Fig. 4.3c). These three different shapes of the slug are essentially due to the balance between the buoyancy and inertia forces. The shape of these slugs also affects their translational velocity through the pipe. Based on the visual observations it is seen that the velocity of downward pointing slug is relatively greater than the velocity of a upward pointing blunt nose shaped slug and flat headed slug. Literature reports that these different forms of vertical downward slug flow exist only for a certain range of pipe diameter approximately for 10 mm $< D <$ 50 mm. It should be noted that the different motions and shapes of gas slug may affect the two phase pressure drop and heat transfer characteristics and hence it is very important to understand the transition (in terms of the phase velocity known as bubble turning velocity) between these different forms of slug flows. However, literature lacks reliable methods to provide information on this transition as a function of fluid combination and pipe diameter.

The froth and churn flow regimes in vertical upward flow has no definitive physical structure and hence their flow visualization is not reported here. The falling film flow in vertical downward flow occurs at low liquid and moderate gas flow rates and is characterized by the flow of liquid gliding smoothly over the pipe surface while the gas phase flows through the core. This type of flow is mostly encountered in film type vertical downward condensers. A major precaution that needs to be taken while operating in the non-isothermal falling film flow regime is

Fig. 4.4 Two phase annular flow in vertical upward pipe orientation

to avoid dry spots on the pipe wall due to excess amount of heat transfer that may further lead to system damage. Since there is no major difference in the physical structure of the falling film flow even with increase in the gas and liquid flow rates, its flow visualization is not reported. The annular flow regime in both vertical upward and downward pipe orientations can be described as the flow of turbulent gas core inside a wavy liquid annulus in contact with the pipe wall. In case of vertical upward and downward flow, there is symmetry about the pipe axis and hence the film thickness can be assumed to be the same throughout the pipe circumference. Whereas, in case of horizontal flow, the flow nature is asymmetric about the pipe axis and the liquid film thickness near the pipe bottom wall is greater than the pipe upper wall due to gravity acting on liquid phase. The gas-liquid interface is wavy and continuous but is dynamic in nature. The liquid film thickness is found to be a function of flow rates, pipe diameter, and orientation and fluid properties. As shown in Figs. 4.4(a–c) and 4.5(a–c), for a fixed pipe diameter ($D = 12.5$ mm) and fluid combination (air-water), the liquid film thickness is observed to decrease with increase in the gas flow rate due the shearing action of gas phase on continuous liquid phase. The annular two phase flow is studied extensively in the literature due to enhanced pressure drop and heat transfer characteristics in this flow regime.

The sequence and the transition of above discussed key flow patterns in horizontal, vertical upward and downward pipe orientations are presented in terms of flow pattern maps in Figs. 4.6, 4.7 and 4.8, respectively. In case of horizontal flow, bubbly flow appears for high liquid and low gas flow rates. Plug flow appears at relatively moderate liquid and low gas flow rates and with increase in the gas flow rate, the flow pattern shifts to slug, wavy slug and finally to annular flow regime. Similar sequence of flow patterns appearance is found in case of vertical upward

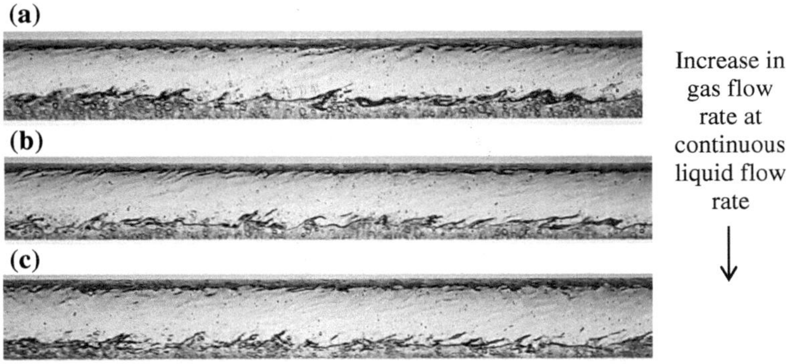

Fig. 4.5 Two phase annular flow in horizontal pipe orientation

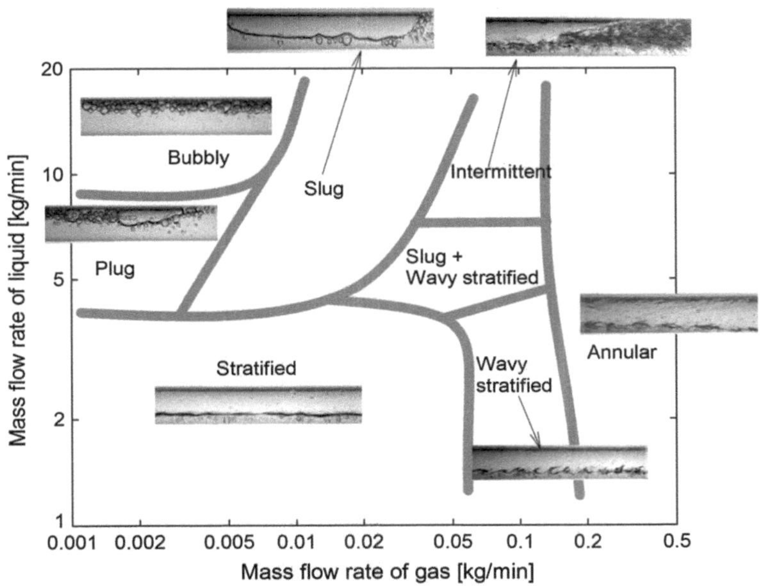

Fig. 4.6 Flow pattern map for horizontal flow

and downward pipe orientations. However, a noticeable difference in the transition boundaries is observed among these three orientations. In comparison to vertical upward and downward flow, the horizontal bubbly flow occupies a small region on the flow pattern map. It appears at comparatively high liquid mass flow rates and transits to slug flow quickly at the maximum gas mass flow rate of 0.1 kg/min. In case of vertical upward orientation, the slug flow exists at smaller liquid and gas flow rates compared to vertical downward and horizontal pipe orientations. This is probably due to the fact that, the net upward flow assists the buoyancy of the gas slug and thus this flow pattern requires smaller gas flow rates to appear.

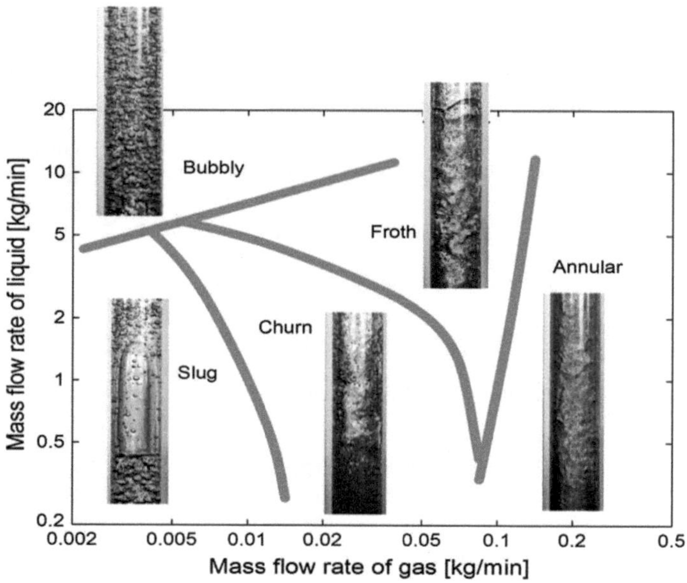

Fig. 4.7 Flow pattern map for vertical upward flow

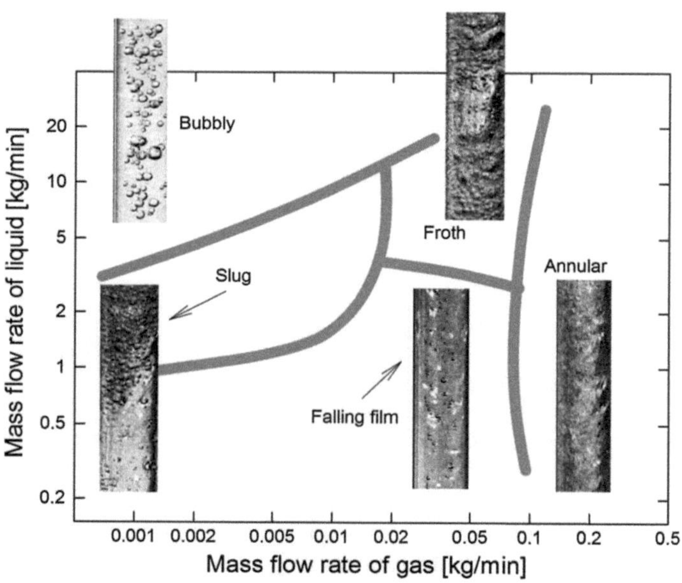

Fig. 4.8 Flow pattern map for vertical downward flow

Although some divergence is observed for the churn annular transition line at low liquid mass flow rates in vertical upward flow; the transition line that separates the annular flow pattern from other flow patterns has approximately the same trend for vertical upward, downward and horizontal pipe orientations. Stratified flow exists for horizontal, downward inclined and near horizontal upward inclined pipe orientations. Experimental flow visualization work of [42] has shown that the transition line in horizontal flow pattern map that separates stratified flow from other flow patterns go away with increase in upward inclined pipe orientations while there is a little effect on transition between other flow patterns with change in pipe orientation from 0° to $+7^{\circ}$ from horizontal.

It should be noted that the flow pattern maps reported here are specific to the experimental setup at Two Phase Flow Lab, Oklahoma State University and the transition boundaries between different flow patterns may alter with change in fluid properties, pipe diameter and pipe orientation. Also as mentioned earlier, since there is no reliable quantitative tool available in the literature to predict the existence and transition of one flow pattern to another for varying two phase flow conditions it is always recommended to use the void fraction and pressure drop correlations that are independent of the flow pattern but still account for the relevant two phase flow physics.

4.3 Void Fraction in Gas-Liquid Two Phase Pipe Flow

As mentioned earlier, the void fraction in gas liquid two phase flow is an indispensable parameter required in several calculations such as the estimation of two phase mixture density required in calculation of two phase hydrostatic pressure drop and the refrigerant charge inventory. The knowledge of void fraction is also crucial in thermo-hydraulic simulations and in determination of two phase natural circulation loop flow rates and the corresponding heat transfer rates. Void fraction in gas-liquid two phase flow is found to be a function of several parameters such as the flow patterns, fluid properties and pipe diameter and orientation. In comparison to the effect of fluid properties and pipe diameter, the effect of flow patterns and pipe orientation on the void fraction is more prominent. The purpose of this section is to report the relationship between void fraction, flow patterns and pipe orientation.

4.3.1 Effect of Flow Patterns (Phase Flow Rates) on Void Fraction

The void fraction in gas-liquid two phase flow is sensitive to the flow patterns. Irrespective of the pipe diameter, orientation and fluid properties, the void fraction increases sharply with a small increase in the gas flow rate for bubbly and slug flow regimes. Whereas, in case of the annular flow, the void fraction is observed to

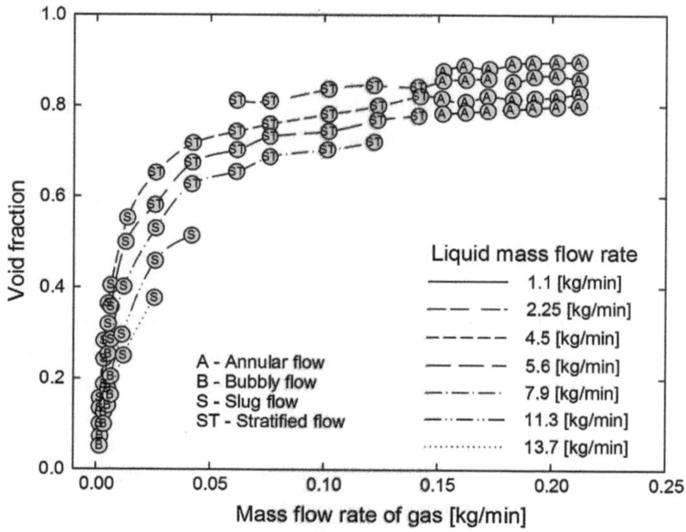

Fig. 4.9 Variation of void fraction with change in the gas and liquid flow rates for different flow patterns in horizontal flow

remain nearly constant even with a major increase in the gas flow rate. The general trend of the increase in void fraction with increase in gas flow rate for various flow patterns in non-boiling horizontal, vertical upward and vertical downward two phase flow is shown in Figs. 4.9, 4.10 and 4.11, respectively. The experimental data is collected for air-water fluid combination in a 12.5 mm I.D. polycarbonate pipe at Two Phase Flow Lab, Oklahoma State University.

4.3.2 Effect of Pipe Orientation on Void Fraction

Regardless of whether or not the flow is boiling or non-boiling in nature, the void fraction in gas-liquid two phase flow is observed to be a strong function of the pipe orientation. In comparison to horizontal and vertical pipe orientations the experimental work related to void fraction and pressure drop in upward and downward inclined pipe orientations is limited. Some of the experimental work available in the literature for inclined two phase flow conditions are those of [10, 74, 76] for non-boiling flow and that of [65] for boiling two phase flow. All of these studies report a noticeable effect of the pipe orientation on void fraction. As shown in Fig. 4.12, for fixed flow conditions, the void fraction first decreases and then increases with increase in the pipe inclination in upward direction. In case of downward inclinations measured from horizontal, the void fraction first increases and later decreases with increase in the pipe orientation. The void fraction is maximum for vertical downward flow in comparison to horizontal and vertical

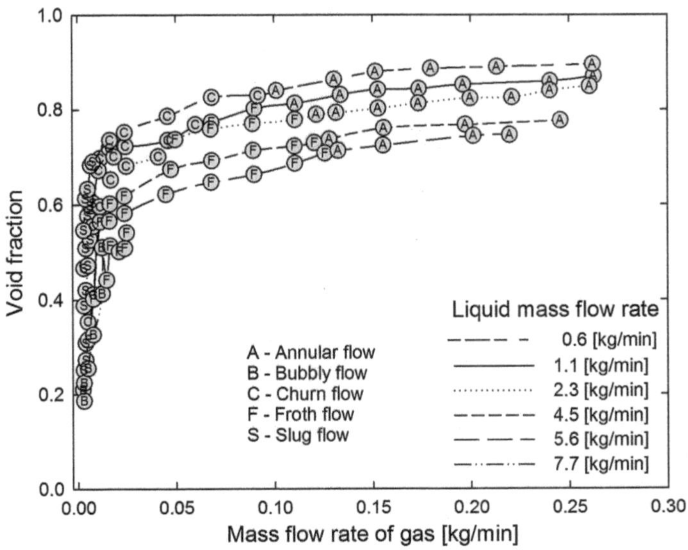

Fig. 4.10 Variation of void fraction with change in the gas and liquid flow rates in vertical upward flow

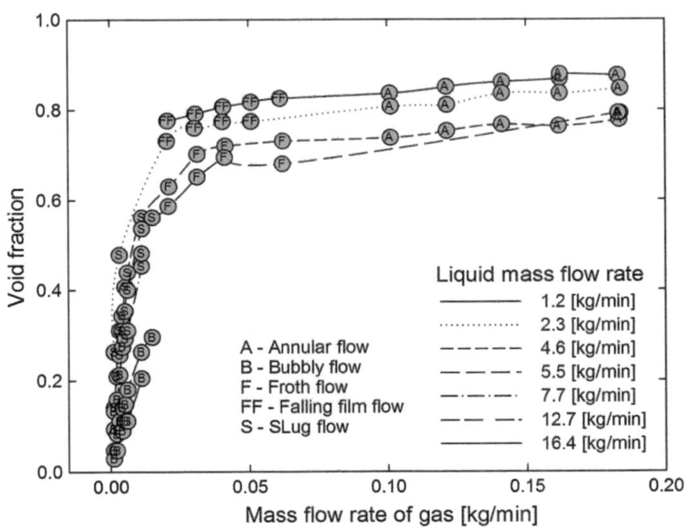

Fig. 4.11 Variation of void fraction with change in the gas and liquid flow rates in vertical downward flow

upward flow. In case of boiling two phase flow, the minima in void fraction is observed to shift towards lower inclination from horizontal with increase in the two phase mixture quality. Another interesting trend observed for boiling two

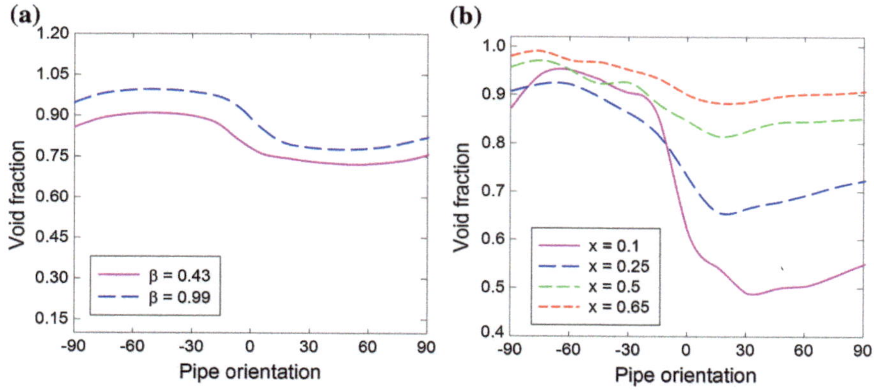

Fig. 4.12 Effect of pipe orientation on void fraction. **a** Non boiling flow (air-water) data from [10]. **b** Condensing flow (R134a) data from [65]

phase flow is that, with increase in the two phase quality, the effect of pipe orientation on void fraction diminishes and thus it is expected that for very high qualities, void fraction remain virtually unaffected by the pipe orientation.

This effect of the pipe orientation on the void fraction can be explained with the help of the concept of residence time of the gas phase inside a given pipe section. For instance, when the void fraction is measured using quick closing valves (QCV) then, the amount of gas phase trapped in the given pipe section depends upon the relative magnitudes of buoyancy, inertia and gravity forces. In case of vertical downward flow, the buoyancy and inertia forces are opposite in direction that opposes the mean flow and hence results into higher residence time of the gas phase and hence higher void fraction. In contrast to this, the residence time of the gas phase in the upward flow is lower where the buoyancy complements the mean direction of two phase flow and hence results in lower values of void fraction. Above discussion infers that the void fraction in downward inclined flow is higher than that in upward inclined two phase flow. From Fig. 4.12, it can also be concluded that the relative magnitude of buoyancy and inertia forces change at certain inclinations from horizontal and that causes shift in the trend of void fraction at these pipe orientations. Since, the effect of buoyancy forces will go away with decrease in the pipe diameter, it can be anticipated that the effect of pipe orientation at fixed flow conditions is more prominent for large pipe diameters and is negligible for very small pipe diameters.

4.3.3 Effect of Pipe Diameter on Void Fraction

The effect of pipe diameter on gas-liquid void fraction is not as significant as the flow patterns or the pipe orientation. The effect of pipe diameter on void fraction for flow of R410A in horizontal pipe orientation is shown in Fig. 4.13. It is evident

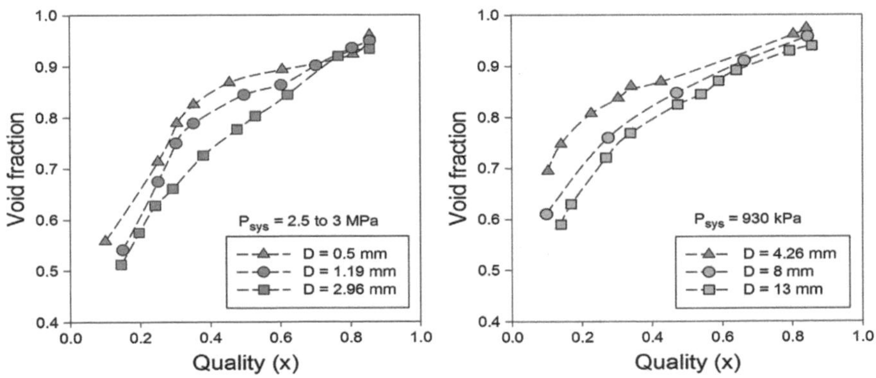

Fig. 4.13 Effect of pipe diameter on void fraction in horizontal two phase flow of R410A

that the void fraction increases with decrease in the pipe diameter for low values of quality or typically in bubbly and slug flow regimes. The effect of pipe diameter is observed to vanish gradually with increase in the two phase flow quality. This observation is in agreement with the work of [55] who investigated the effect of pipe diameter in a range of 5 mm < D < 50 mm on void fraction in vertical upward annular flow regime for non-boiling two phase flow. They found that the effect of pipe diameter on void fraction in the annular flow regime is negligible. Due to lack of experimental data for similar fluid combinations the effect of pipe diameter on two phase pressure drop for various pipe orientations could not be analyzed. However, based on the trend of the data in Fig. 4.13 for horizontal two phase flow and the results of [55] for vertical upward flow, it can be conjectured that the effect of pipe diameter on void fraction is independent of the pipe orientation for annular flow regimes. More experimental data and in-depth analysis is required to verify this effect for other flow patterns.

From the discussion so far, it is evident that void fraction is influenced by the flow patterns, pipe orientation and pipe diameter. Thus, for successful prediction of the void fraction it is very much desired for any correlation to account for the effect of pipe orientation, pipe diameter and flow patterns on the void fraction.

4.4 Performance Assessment of Two Phase Void Fraction Correlations

Since last few decades, significant contribution has been made by the scientific community to experimental measurements and modeling of void fraction in gas-liquid two phase flow. The initial experimental and modeling work in this field was mostly focused on the air-water two phase flow and later got diversified by the inclusion of boiling two phase flow phenomenon encountered in refrigeration and nuclear industry and the simultaneous flow of oil-gas commonly observed in petroleum industry. In spite of the numerous void fraction correlations available in the literature, there is no single correlation that can be relied upon to predict the void fraction correctly for a wide range of two phase flow situations. Most of these available correlations have limited application due to the restrictions imposed on them for certain fluid combinations, pipe orientations and pipe diameters. As a result, the main problem for a design engineer is to make a correct choice of the void fraction correlation for the desired application. The objective of this section is to revisit some of the renowned and widely used void fraction correlations and recommend the rigorously scrutinized top performing correlations for air-water and refrigerant two phase flow for a range of pipe diameters and pipe orientations.

The void fraction correlations reported in the literature can be broadly classified in three main categories. For instance, the correlations based on the concept of the slip (separated flow) model, the drift flux model and empirical correlations. The slip (separated flow) model based correlations assume that the two phases flow separately with different velocities and share a definite interface. The flow patterns such as stratified and annular flow behave as a separated flow and can be effectively modeled using these types of correlations. The slip model based correlations are mostly preferred in refrigeration industry due to the stratified and annular flow pattern dominated two phase flow in evaporators and condensers. The category of drift flux model based correlations assumes one phase dispersed in other continuous phase and requires the determination of distribution parameter and drift velocity as variables to calculate void fraction. The flow patterns such as bubbly flow, slug flow and mist flow are the preferred flow patterns to be modeled using the concept of drift flux. The application of empirical correlations is mostly restricted by the range of experimental data used to determine the constants and hence tend to fail if used for a wide range of flow conditions out of the scope of that particular correlation.

Let's first consider some of the widely used separated flow model based void fraction correlations reported in the literature. The general form of the separated flow model may be expressed as shown in Eq. (4.1).

$$\alpha = \frac{1}{1 + \left(\frac{1-x}{x}\right)\left(\frac{\rho_g}{\rho_l}\right)S} \tag{4.1}$$

The slip ratio (S) in Eq. (4.1) is the ratio of actual cross sectional averaged velocities of the gas and liquid phase as expressed by Eq. (4.2).

$$S = \frac{U_g}{U_l} = \frac{(U_{sg}/\alpha)}{(U_{sl}/(1-\alpha))} \tag{4.2}$$

With the homogeneous flow model assumption of equal velocities of both gas and liquid phase, the slip ratio (S) in above equation is equal to unity. However, in reality $S \geq 1$ or $S < 1$ depending upon the pipe orientation and hence the above equation is modified by different investigators to account for the slippage between gas and liquid phase. The assumption of homogeneous flow remains valid only for certain region of two phase flow where $S \approx 1$ as in case of very small region of

Table 4.1 Void fraction correlations based on slip (separated) flow model

Correlation	A	p	q	r
homogeneous	1	1	1	0
[114]	1	1	0.67	0
[101]	1	0.72	0.4	0.08
[66]	0.28	0.64	0.36	0.07
[98]	1	1	0.89	0.18
[7]	1	0.74	0.65	0.13
[93]	2.22	0.65	0.65	0
[19]	0.18	0.6	0.33	0.07
[113]	$\alpha^{-0.125}$	0.875	0.875	0.875
[91]	$0.4+0.6\sqrt{\left(\frac{\rho_L}{\rho_g}+0.4\left(\frac{1-x}{x}\right)\right)/\left(1+0.4\left(\frac{1-x}{x}\right)\right)}$	1	1	0
[83]	$1+1.578\mathrm{Re}_{sl}^{-0.19}\left(\frac{\rho_L}{\rho_g}\right)^{0.22}\left(\frac{y}{1+yF_2}-yF_2\right)$ $y=\left(\frac{(1-x)}{x}\frac{\rho_g}{\rho_l}\right)^{-1}$, $F_2=0.0273We_{sl}\mathrm{Re}_{sl}^{-0.51}\left(\frac{\rho_L}{\rho_g}\right)^{-0.08}$	1	1	0
[20]	$\sqrt{1-x\left(\frac{\rho_l-\rho_g}{\rho_l}\right)}$	1	1	0

void fraction (bubbly flow) or very large region of void fraction (annular mist flow). The general form of separated flow models that account for the slip between the two phases may be expressed by rewriting Eq. (4.1) as shown in Eq. (4.3). The different correlations are listed in Table 4.1.

$$\alpha=\frac{1}{1+A\left(\frac{1-x}{x}\right)^p\left(\frac{\rho_g}{\rho_l}\right)^q\left(\frac{\mu_l}{\mu_g}\right)^r} \tag{4.3}$$

The void fraction correlations based on drift flux model expressed by Eq. (4.4) require determination of distribution parameter (C_o) and drift velocity (U_{gm}) as listed in Table 4.2.

$$\alpha=\frac{U_{sg}}{C_oU_m+U_{gm}}\ \left(\text{where } U_m=U_{sg}+U_{sl}\right) \tag{4.4}$$

The performance of the void fraction correlations based on separated flow model and drift flux model considered in Tables 4.1 and 4.2 is assessed against a comprehensive data of air water and liquid refrigerant and its vapor consisting of 5162 (27 sources) and 645 (8 sources) data points, respectively. The experimental data used for performance assessment is summarized in Tables 4.3 and 4.4. The air water data consists of horizontal, upward and downward inclined pipe orientations while the refrigerant liquid vapor data consists of horizontal pipe orientation only.

The performance of separated flow model and drift flux model based correlations is assessed based on the percentage of data points predicted within certain error bands criterion for four different ranges of the void fraction namely, $0<\alpha\leq 0.25$, $0.25<\alpha\leq 0.5$, $0.5<\alpha\leq 0.75$ and $0.75<\alpha<1$, respectively. These four ranges

Table 4.2 Void fraction correlations based on drift flux model

Correlation	Distribution parameter	Drift velocity (m/s)
[12]	$C_o = \left[\frac{1}{(1+\cos\theta)^{1.25}}\right]^{(1-\alpha)^{0.5}} + 0.18\left(\frac{U_{sl}}{U_m}\right)^{0.1}$	$U_{gm} = (\mu_l/\mu_w)^{-0.25}(0.35\sin\theta + 0.54\cos\theta)\sqrt{gD(\Delta\rho/\rho_l)}(1-\alpha)^{-0.5\sin\theta}$
[105]	$C_o = \frac{U_{sg}}{U_m}\left(1 + \left(\frac{U_{sl}}{U_{sg}}\right)^{b_1}\right)$ where $b_1 = \left(\frac{\rho_g}{\rho_l}\right)^{0.1}$	$U_{gm} = 2.9\left[gD\sigma(1+\cos\theta)\Delta\rho/\rho_l^2\right]^{0.25}(1.22 + 1.22\sin\theta)^{b_2}$ where $b_2 = (P_{atm}/P_{sys})$ The leading constant of 2.9 carries units of $m^{-0.25}$
[86]	$C_o = 1 + 0.2(1-x)\left(gD\rho_l^2/G^2\right)^{0.25}$ $C_o = 1 + 0.12(1-x)$ (for horizontal flow)	$U_{gm} = 1.18(1-x)\left(\sigma g\Delta\rho/\rho_l^2\right)^{0.25}$
[45]	$C_o = 1.15$	$U_{gm} = 1.53\left(g\sigma\Delta\rho/\rho_l^2\right)^{0.25}(1-\alpha)^{0.5}\sin\theta$
[15]	$C_o = 1.2$	$U_{gm} = 0.35\sqrt{gD(\Delta\rho/\rho_l)}$
[47]	$C_o = 1$	$U_{gm} = 0.671\sqrt{gD}(\sin\theta)^{0.263}$
[52][a]	For $0 \le \left(U_{sg}^+/U_m^+\right) \le 0.9$ where $U_{sg}^+ = U_{sg}\Big/\left(\sigma g\Delta\rho/\rho_l^2\right)^{0.25}$, $U_m^+ = U_m\Big/\left(\sigma g\Delta\rho/\rho_l^2\right)^{0.25}$ $C_o = \exp\left\{0.475\left(\frac{U_{sg}}{U_m}\right)^{1.69}\right\}\left(1 - \sqrt{\frac{\rho_g}{\rho_l}}\right) + \sqrt{\frac{\rho_g}{\rho_l}}$ Else $C_o = \left\{-2.88\left(\frac{U_{sg}}{U_m}\right) + 4.08\right\}\left(1 - \sqrt{\frac{\rho_g}{\rho_l}}\right) + \sqrt{\frac{\rho_g}{\rho_l}}$	$U_{gm} = \sqrt{2}\left(\frac{\sigma g\Delta\rho}{\rho_l^2}\right)^{0.25}(1-\alpha)^{1.75}\exp\left(-1.39U_{sg}\left(\frac{\sigma g\Delta\rho}{\rho_l^2}\right)^{-0.25}\right)$ $+ (U_{gm})_{KI}\left\{1 - \exp\left(-1.39U_{sg}\left(\frac{\sigma g\Delta\rho}{\rho_l^2}\right)^{-0.25}\right)\right\}$ $(U_{gm})_{KI}$ is the drift velocity given by [56] (Recommended for large pipe diameters in bubbly flow regime)
[44]	For $-20 \le U_{gm}^* < 0$ where $U_{gm}^* = (U_{gm}/U_m)$ $C_o = \left(-0.0214U_{gm}^* + 0.772\right) + \left(0.0214U_{gm}^* + 0.228\right)\sqrt{\rho_g/\rho_l}$ Else, $C_o = \left(0.2\exp(0.00848U_{gm}^* + 20) + 1\right) - 0.2\exp(0.00848U_{gm}^* + 20)\sqrt{\rho_g/\rho_l}$	$U_{gm} = \sqrt{2}\left(g\sigma\Delta\rho/\rho_l^2\right)^{0.25}$

(continued)

Table 4.2 (continued)

Correlation	Distribution parameter	Drift velocity (m/s)
[57]	For $0 < U_m \leq -3.5$ $C_o = 1.2 - 0.2\sqrt{(\rho_g/\rho_l)}$ For $-3.5 < U_m \leq -2.5$ $C_o = 0.9 + 0.1\sqrt{\rho_g/\rho_l} - 0.3\left(1 - \sqrt{\rho_g/\rho_l}\right)(2.5 + U_m)$ For $-2.5 < U_m < 0$ $C_o = 0.9 + 0.1\sqrt{\rho_g/\rho_l}$	Refer to [57] to determine drift velocity based on several criteria of mixture velocity, system pressure and pipe diameter
[17]	$C_o = 1.185$ (bubbly flow) $C_o = 1.15$ (slug flow)	$U_{gm} = 1.53\left(\sigma g \Delta\rho / \rho_l^2\right)^{0.25}$ $U_{gm} = 0.35\sqrt{gD(\Delta\rho/\rho_l)}$
[26]	$C_o = 0.934(1 + 1.42\alpha)$	$U_{gm} = 1.53(\sigma g \Delta\rho/\rho_l)^{0.25}$
[49]	$C_o = 1.12$	$U_{gm} = 0.345\sqrt{gD(\Delta\rho/\rho_l)}$
[53]	$C_o = 1 + 0.796\exp\left(-0.061\sqrt{\rho_l/\rho_g}\right)$	$U_{gm} = 0.034\left(\sqrt{\rho_l/\rho_g} - 1\right)$
Toshiba[b]	$C_o = 1.08$	0.45
[11]	$C_o = 1$	$U_{gm} = 0.188(gD\Delta\rho/\rho_l)^{0.5}$
[97]	$C_o = 1.11775 + 0.45881\alpha - 0.57656\alpha^2$	$U_{gm} = k\frac{C_o(1-C_o\alpha)}{m^2 + C_o\alpha\left(\sqrt{\rho_g/\rho_l - m^2}\right)}\sqrt{\frac{gD\Delta\rho}{\rho_l}}$ $k = \sqrt{Ku^2/D^*}$, m = 1.36, $Ku = \sqrt{\left(D^*\min\left(\frac{1}{2.4}, \frac{10.24}{D^*}\right)\right)}$, $D^* = D\sqrt{g\Delta\rho/\sigma}$
[56]	For circular pipes $C_o = 1.2 - 0.2\sqrt{(\rho_g/\rho_l)}$ For rectangular pipes $C_o = 1.35 - 0.35\sqrt{(\rho_g/\rho_l)}$	For $N_{\mu f} \leq 0.00225$ and $D^* \leq 30$ $U_{gm} = 0.0019(D^*)^{0.809}(\rho_g/\rho_l)^{-0.157} N_{\mu f}^{-0.562}(g\sigma\Delta\rho/\rho_l)^{0.25}$ For $D^* \geq 30$, $U_{gm} = 0.03(\rho_g/\rho_l)^{-0.157} N_{\mu f}^{-0.562}(g\sigma\Delta\rho/\rho_l)^{0.25}$ For $N_{\mu f} > 0.00225$, $U_{gm} = 0.92(\rho_g/\rho_l)^{-0.157}(g\sigma\Delta\rho/\rho_l)^{0.25}$ $N_{\mu f} = \left(\mu_l / \left(\rho_l \sigma\sqrt{\sigma/g\Delta\rho}\right)^{0.5}\right)$, $D^* = D/\sqrt{\sigma/g\Delta\rho}$

(continued)

Table 4.2 (continued)

Correlation	Distribution parameter	Drift velocity (m/s)
[90]	$C_o = 1.2$	$U_{gm} = \left\{0.24 + 0.35\left(U_{sg}/U_m\right)^2\sqrt{(gD\alpha)}\right\}$
[95]	$C_o = (0.82 + 0.18\ P_{sys}/P_{crit})^{-1}$	$U_{gm} = \sqrt{2}\left(g\sigma\Delta\rho/\rho_l^2\right)^{0.25}$
[70]	$C_o = 1.3$	$U_{gm} = 0.7$
[59]	$C_o = 1.2$	$U_{gm} = 0.345\sqrt{gD(\Delta\rho/\rho_l)}$
[15]	$C_o = 1.2$	$U_{gm} = \pm 0.35\sqrt{gD(\Delta\rho/\rho_l)}$
[112]	$C_o = 1.2 + 0.38\ \exp\ (-1.39/La)$	Drift velocity assumed negligible in mini-channels
[21]	$C_o = \frac{2}{1+\left(Re_{sl}/1000\right)^2} + \frac{1.2-0.2\sqrt{\rho_g/\rho_l}(1-\exp(-18\alpha))}{1+\left(1000/Re_{sl}\right)^2}$	$U_{gm} = 0.0246\cos\theta + 1.606\left(\frac{g\sigma\Delta\rho}{\rho_l}\right)^{0.25}\sin\theta$
[25][c]	$\alpha = \frac{hx^n}{1+(h-1)x^n}$ where, $h = -2.129 + 3.129\left(\rho^g\rho_l^{-1}\right)^{-0.2186}$ and $n = 0.3487 + 0.6513\left(\rho_g\rho_l^{-1}\right)^{0.5150}$	

[a] Refer to the original paper for other flow patterns specific correlation each for bubbly, slug, froth, churn and annular flow regime
[b] Reported by Woldesemayat and Ghajar [105]
[c] Not a drift flux model correlation but considered in this chapter since it performs well in high region of void fraction

Table 4.3 Air-water experimental data used for assessment of void fraction correlations

Data source	D [mm]	Orientation [deg.]	No. of data points	Void fraction range
Two Phase Flow Lab, Oklahoma State University	12.7	0, ±90	530	0.02–0.93
[79]	5	90	58	0.17–0.96
[55]	19	90	57	0.59–0.96
[2]	50	0	56	0.2–0.69
[31]	–	90	61	0.09–0.92
[32]	60.1	0.5, 1, 3	91	0.35–0.97
[81]	51	1, 0, −1, −2	112	0.9–0.98
[5]	78	0	36	0.94–0.99
[22]	75.9	90	103	0.14–0.93
[35]	50.7	90	90	0.03–0.95
[10]	25.4, 38.1	±90, ±85, ±75, ±35, ±20, ±15, ±10, ±5, 0	565	0.09–0.99
[76]	44.5	0, 2.75, 20.75, 45, 70, 90, −6.75, −20.5, −44.5, −67.5, −90	1521	0.01–0.99
[80]	25.4	±90	290	0.05–0.96
[77]	44.5	±90	159	0.05–0.92
[29]	32	−90	26	0.03–0.29
[82]	19	−90	35	0.11–0.95
[50]	50.8	−90	39	0.75–0.97
[110]	9.52	±90	163	0.06–0.97
[67]	44, 90	−90	115	0.01–0.21
[102]	16	−90	25	0.07–0.89
[94]	12.7	90	94	0.01–0.98
[61]	48.2	30, 50, 60, 70, 90	460	0.02–0.82
[37]	19	0	88	0.06–0.94
[72]	77.9	0	54	0.54–0.99
[33]	50.8, 149	0	40	0.25–0.91
[106]	4.5	0, 30, 60	123	0.008–0.96
[3]	25.4	90	171	0.6–0.97

Table 4.4 Refrigerant void fraction data used for assessment of void fraction correlations

Data source	D [mm]	Refrigerant	No. of data points	Void fraction range
[58]	8	R410A	77	0.21–0.99
[89]	0.5, 1.19, 2.96	R410A	142	0.09–0.98
[60]	7.52	R134a	29	0.2–0.98
[109]	4.26	R134a, R410A	36	0.57–0.97
[104]	13.6	R22, R410A	43	0.31–0.98
[16]	7.2	R134a	81	0.35–0.82
[46]	7.04	R410A, R134a	40	0.31–0.95
[87]	9.57	R11, R12, R22	197	0.35–0.99

of void fraction are selected such that they approximately represent different gas-liquid two phase flow patterns. For example, the lowest range of the void fraction approximately represents the bubbly flow regime, the intermediate two ranges of the void fraction represent slug, churn, froth and stratified flow regimes while the last range of void fraction represents annular flow regime. The knowledge of void fraction in two phase flow is usually of no interest unless it is used to calculate other quantities required for process design and sizing of equipment. For example, one of the important uses of void fraction is to calculate the hydrostatic pressure drop based on the correct estimation of two phase mixture density which in turn depends on the correct prediction of the void fraction. Thus the selection of void fraction largely depends upon the desired accuracy or the acceptable error associated with the derived quantities. Ghajar and Bhagwat [41] showed that the void fraction correlation directly influences the two phase mixture density and this influence is of different magnitude for different ranges of the void fraction. They reported that an error of up to ±30 % in small range of void fraction ($0 < \alpha \leq 0.25$) has a small effect on two phase mixture density while a relatively small error of ±10 % in prediction of the large range of void fraction ($0.75 < \alpha < 1$) may cause a considerable error in estimation of the two phase mixture density. Taking into account this relation between error in prediction of the void fraction and its effect of two phase mixture density, an error of up to ±30 % is accepted for the low range of void fraction while an error less than ±10 % is considered acceptable for the highest range of the void fraction. For the two intermediate ranges of the void fraction, an error up to ±20 % is considered acceptable. In past, the work of [43] shortlisted top performing correlations for horizontal, vertical upward and vertical downward flow based on a similar criterion. Their recommendations were based on the percentage of data points predicted within ±30, ±20, ±15 and ±10 % error bands for four different ranges of the void fraction, respectively. The void fraction data base for air water fluid combination consists of several pipe orientations and hence the assessment of the void fraction correlations is carried out for the above mentioned ranges of the void fraction for seven different classifications of the pipe orientations as shown in Tables 4.5, 4.6, 4.7 and 4.8. In order to select the appropriate correlation from these tables, users may have to solve the top performing correlations on an iterative basis using computer program. To simplify this process, for each void fraction range, corresponding gas volumetric flow fraction range (β) is also reported in these tables. The gas volumetric flow fraction (β) defined as the ratio of the superficial velocity of the gas phase to the total two phase (mixture) velocity is known based on the flow rate of each phase. Thus, it is recommended for users to first verify the gas volumetric flow fraction range and then use the appropriate correlation for corresponding void fraction range.

As shown in Table 4.5, for the first two ranges of void fraction in vertical downward pipe orientation, the correlation of [12] predicts the highest number of data points within ±30 and ±20 % error bands. For the next two ranges of void fraction i.e., $0.5 < \alpha \leq 0.75$ and $0.75 < \alpha < 1$, the correlations of [25, 105] give better accuracy than [12] correlation. As pointed out earlier, the correlation of [25] is based on a modeling method in biochemical kinetics and is neither a separated

Table 4.5 Top performing correlations for vertical upward and downward flow

Void fraction range	$\theta = -90^o$		Void fraction range	$\theta = 90^o$	
$0 < \alpha \leq 0.25$ (248 data points)	±30 %		$0 < \alpha \leq 0.25$ (294 data points)	±30 %	
$0 < \beta \leq 0.25$			$0 < \beta \leq 0.45$		
[12]	96		[12]	84	
[21]	96		[86]	89	
[45]	93		[90]	88	
$0.25 < \alpha \leq 0.5$ (127 data points)	±20 %		$0.25 < \alpha \leq 0.5$ (253 data points)	±20 %	
$0.2 < \beta \leq 0.55$			$0.3 < \beta \leq 0.8$		
[12]	80		[90]	85	
[21]	76		[12]	84	
[20]	76		[52][a]	83	
$0.5 < \alpha \leq 0.75$ (130 data points)	±20 %		$0.5 < \alpha \leq 0.75$ (363 data points)	±20 %	
$0.4 < \beta \leq 0.95$			$0.5 < \beta \leq 0.95$		
[12]	76		[52][1]	95	
[105]	82		[45]	93	
[25]	88		[12]	92	
$0.75 < \alpha < 1$ (370 data points)	±5 %	±10 %	$0.75 < \alpha < 1$ (517 data points)	±5 %	±10 %
$0.65 < \beta < 1$			$0.85 < \beta < 1$		
[12]	51	86	[91]	65	88
[105]	62	82	[105]	63	89
[25]	58	76	[25]	70	91

[a] Flow pattern specific correlation for slug flow

flow nor a drift flux model based correlation whereas [105] is a drift flux model based void fraction correlation. For the lowest range of void fraction in vertical upward flow, the correlation of [86] gives best performance and predicts 89 % of data points within ±30 % error bands. The correlation of [12] also gives comparable performance for the first three ranges of void fraction i.e., for $\alpha \leq 0.75$. Similar to vertical upward flow, for the last range of void fraction in vertical upward flow [25, 105] are among the top performers.

Based on the pipe orientation, the experimental void fraction data in downward inclined pipe orientations is divided into two categories namely $0^o > \theta \geq -45^o$ and $-45^o > \theta > -90^o$. Due to very limited data available in the literature, the entire range of void fraction is divided into three categories i.e., $0 < \alpha \leq 0.5$, $0.5 < \alpha \leq 0.75$ and $0.75 < \alpha < 1$. As listed in Table 4.6, for $0^o > \theta \geq -45^o$, correlation of [12] is among the top three performing correlations for $\alpha \leq 0.75$. For the large values of void fraction, the correlations of [25, 91, 105] are the top performers. In case of pipe orientation in a range of $-45^o > \theta > -90^o$, correlation of [105] consistently stays among the top three performing correlations for every range of void fraction.

Similar to downward inclined pipe orientations, the top performing correlations for the void fraction data in upward inclined pipe orientations are also categorized

Table 4.6 Top performing correlations for downward inclined pipe orientations

Void fraction range	$0^{\circ} > \theta \geq -45^{\circ}$		Void fraction range	$-45^{\circ} > \theta > -90^{\circ}$	
$0 < \alpha \leq 0.5$ (33 data points)	±30 %		$0 < \alpha \leq 0.5$ (15 data points)	±30 %	
$0 < \beta \leq 0.5$			$0 < \beta \leq 0.4$		
[83]	79		[105]	75	
[12]	76		[12]	63	
[57]	70		[114]	75	
$0.5 < \alpha \leq 0.75$ (59 data points)	±20 %		$0.5 < \alpha \leq 0.75$ (18 data points)	±20 %	
$0.4 < \beta \leq 0.9$			$0.3 < \beta < 0.8$		
[25]	93		[105]	72	
[12]	90		[25]	72	
[114]	88		[83]	72	
$0.75 < \alpha < 1$ (368 data points)	±5 %	±10 %	$0.75 < \alpha < 1$ (89 data points)	±5 %	±10 %
$0.8 < \beta < 1$			$0.8 < \beta < 1$		
[105]	66	78	[105]	81	88
[25]	69	81	[19]	83	94
[91]	65	79	[91]	79	88

for two different ranges of the pipe orientations i.e., $0^{\circ} < \theta \leq 45^{\circ}$ and $45^{\circ} < \theta < 90^{\circ}$ as shown in Table 4.7. Sufficient void fraction data is available for the upward inclined pipe orientation and hence the top performing correlations are presented for all four ranges of the void fraction as discussed before. For both ranges of pipe orientation in upward inclined flow, the correlation of [12] is consistently a top performer for first three ranges of void fraction whereas for the large values of void fraction ($\alpha > 0.75$) [105] gives the best accuracy.

For both ranges of the pipe orientations i.e., $0^{\circ} < \theta \leq -45^{\circ}$ and $45^{\circ} < \theta < 90^{\circ}$ [12] correlation predicts the highest number of data points for the first three ranges of the void fraction up to $\alpha \leq 0.75$. For the highest range of the void fraction i.e., $0.75 < \alpha < 1$, correlation of [105] is observed to give best performance and is followed by the performance of [91] correlation. In case of horizontal two phase flow, limited data is available for the first two ranges of void fraction. As shown in Table 4.8, the correlation of [83] gives best performance for the first two ranges of the void fraction. For the void fraction range of $0.25 < \alpha \leq 0.5$, correlation of [105] also gives comparable performance to that of [83]. For the next two ranges of the void fraction i.e., $0.5 < \alpha \leq 0.75$ and $0.75 < \alpha < 1$ [105] gives best accuracy by predicting more than 90 % of data points within ±20 and ±10 % error bands, respectively.

In case of refrigerant void fraction data in horizontal two phase flow, the top five performing correlations are presented in Table 4.9. It is observed that the correlation of [105] gives consistent and superior performance in all three ranges of the void fraction. Although the accuracy of [12] correlation is not comparable with the other top performing correlations listed in Table 4.9, it is evident that this correlation gives comparable performance for the first two ranges of the void

Table 4.7 Top performing correlations for upward inclined pipe orientations

Void fraction range	$0^\circ < \theta \le 45^\circ$		Void fraction range	$45^\circ < \theta < 90^\circ$	
$0 < \alpha \le 0.25$ (159 data points)	±30 %		$0 < \alpha \le 0.25$ (283 data points)	±30 %	
$0 < \beta \le 0.85$			$0 < \beta \le 0.5$		
[12]	84		[12]	85	
[47]	67		[47]	84	
[8]	64		[53]	81	
$0.25 < \alpha \le 0.5$ (167 data points)	±20 %		$0.25 < \alpha \le 0.5$ (139 data points)	±20 %	
$0.3 < \beta \le 0.95$			$0.3 < \beta \le 0.8$		
[12]	90		[12]	93	
[105]	87		[47]	83	
[86]	79		[90]	86	
$0.5 < \alpha \le 0.75$ (158 data points)	±20 %		$0.5 < \alpha \le 0.75$ (104 data points)	±20 %	
$0.6 < \beta \le 0.95$			$0.55 < \beta \le 0.95$		
[105]	95		[105]	99	
[12]	94		[12]	100	
[52][a]	98		[90]	100	
$0.75 < \alpha < 1$ (521 data points)	±5 %	±10 %	$0.75 < \alpha < 1$ (148 data points)	±5 %	±10 %
$0.9 < \beta < 1$			$0.9 < \beta < 1$		
[105]	83	92	[105]	89	98
[25]	82	96	[12]	70	91
[91]	79	96	[91]	64	89

[a] Flow pattern specific correlation for slug flow

Table 4.8 Top performing correlations for horizontal pipe orientation

Void fraction range	$\theta = 0^\circ$	
$0 < \alpha \le 0.25$ (33 data points)	±30 %	
$0 < \beta \le 0.25$		
[12]	76	
[91]	79	
[83]	82	
$0.25 < \alpha \le 0.5$ (88 data points)	±20 %	
$0.2 < \beta \le 0.5$		
[105]	90	
[19]	91	
[83]	90	
$0.5 < \alpha \le 0.75$ (183 data points)	±20 %	
$0.5 < \beta \le 0.95$		
[12]	93	
[105]	95	
[86]	95	
$0.75 < \alpha < 1$ (295 data points)	±5 %	±10 %
$0.95 < \beta < 1$		
[105]	83	99
[25]	74	92
[86]	77	97

Table 4.9 Top performing correlations for refrigerant vapor two phase flow in horizontal pipe

Void fraction range	$0 < \alpha \leq 0.5$	$0.5 < \alpha \leq 0.75$		$0.75 < \alpha < 1$	
No. of data points	73	184		388	
Correlations	±30 % error bands	±20 %	±30 %	±5 %	±10 %
[20]	48	57	62	73	87
[91]	51	54	61	78	88
[7]	53	43	60	59	79
[105]	**49**	**60**	**62**	**87**	**91**
[86]	49	54	61	80	88
[12]	45	56	60	28	73

Table 4.10 Recommended void fraction correlations for different two phase flow situations

θ [deg]	D [mm]	α	β	Correlation
Air-water data				
−90	9–90	$0 < \alpha \leq 0.5$	$0 < \beta \leq 0.55$	[12]
		$0.5 < \alpha < 1$	$0.45 < \beta < 1$	[105]
90	5–200	$0 < \alpha \leq 0.75$	$0 < \beta \leq 0.95$	[12]
		$0.75 < \alpha < 1$	$0.85 < \beta < 1$	[105]
$-45 \leq \theta < 0$	25–50	$0 < \alpha \leq 0.75$	$0 < \beta \leq 0.9$	[12]
		$0.75 < \alpha < 1$	$0.8 < \beta < 1$	[105]
$-90 < \theta < -45$	25–45.5	$0 < \alpha < 1$	$0 < \beta < 1$	[105]
$0 < \theta \leq 45$	25–60	$0 < \alpha \leq 0.75$	$0 < \beta \leq 0.95$	[12]
		$0.75 < \alpha < 1$	$0.9 < \beta < 1$	[105]
$45 < \theta < 90$	25–50	$0 < \alpha \leq 0.75$	$0 < \beta \leq 0.95$	[12]
		$0.75 < \alpha < 1$	$0.9 < \beta < 1$	[105]
0	12.5–149	$0 < \alpha \leq 0.25$	$0 < \beta \leq 0.25$	[83]
		$0.25 < \alpha < 1$	$0.2 < \beta < 1$	[105]
Refrigerant data				
0	0.5–13.6	$0 < \alpha < 1$	$0 < \beta < 1$	[105]

fraction. Overall it is found that for refrigerant two phases flow, even the top performing correlations have limited success in prediction of the void fraction for $0 < \alpha \leq 0.5$ and $0.5 < \alpha \leq 0.75$. For the highest range of the void fraction correlation of [105] gives best accuracy and predicts 87 % of data points within ±5 % error bands. Based on the overall performance of the correlations listed in Table 4.9, it is receommended to use [105] correlation independent of the flow patterns and void fraction range.

The top performing correlations reported in Tables 4.5, 4.6, 4.7, and 4.8, can be used for different ranges of void fraction (α) and gas volumetric flow fraction (β). It is seen that some of these top performing correlations perform consistently for most of the two phase flow conditions. In order to avoid selection and use of different correlations for different ranges of pipe orientation and void fraction, following two correlations are recommended for different two phase flow scenarios. As shown in Table 4.10 [12, 105] correlations can be used to predict void fraction for various ranges of pipe orientations and gas volumetric flow fraction. In case of the near vertical downward inclined pipe orientations i.e., $-90^{\circ} < \theta < -45^{\circ}$ (Table 4.6) and two phase flow of refrigerant through horizontal pipes (Table 4.9), the correlation of [105] may be used for the entire range of void fraction and gas volumetric flow fraction. In case the gas volumetric flow fraction overlaps with the two different ranges of the void fraction then it is recommended to calculate the void fraction using both [12, 105] correlations and select the maximum of two values since the [12] correlation comparatively tend to under predict the void fraction. The selection of correct void fraction correlation for such a case is elucidated at the end of this book chapter in form of a solved problem.

4.5 Pressure Drop in Gas-Liquid Two Phase Flow

One of the most common requirements for design and sizing of industrial process and equipment is the determination of pressure drop in the system. This section of the chapter will attempt to provide insight about the different two phase flow variables that affect the two phase frictional pressure drop. In addition to this, this section also provides a brief synopsis of some of the existing methodologies available in the literature and their accuracies associated with the prediction of the two phase pressure drop.

The total pressure drop in gas-liquid two phase flow essentially consists of three components namely, hydrostatic, frictional and accelerational pressure drop. As shown in Eq. (4.5), these three components are additive in nature and in order to calculate total pressure drop, each of these components has to be calculated separately. The contribution of each of these components to the total two phase pressure drop depends on the flow pattern, void fraction, pipe diameter, pipe orientation and the type of two phase flow i.e., boiling or non-boiling two phase flow. The hydrostatic component of the two phase pressure drop is due to the pipe elevation and is calculated using two phase mixture density which in turn depends upon the accurate estimation of void fraction. The hydrostatic component of two phase pressure drop can be calculated as shown in Eq. (4.6). The two phase mixture density based on void fraction is calculated as shown in Eq. (4.7).

$$\left(\frac{dP}{dL}\right)_{t,tp} = \left(\frac{dP}{dL}\right)_h + \left(\frac{dP}{dL}\right)_f + \left(\frac{dP}{dL}\right)_a \tag{4.5}$$

$$\left(\frac{dP}{dL}\right)_h = \rho_{tp} g \sin\theta \tag{4.6}$$

$$\rho_{tp} = (1-\alpha)\rho_l + \alpha\rho_g \tag{4.7}$$

In case if the void fraction is not known, two phase flow literature also reports another simple way of calculating two phase mixture density using flow quality as shown in Eq. (4.8). This approach is called a homogeneous flow model approach

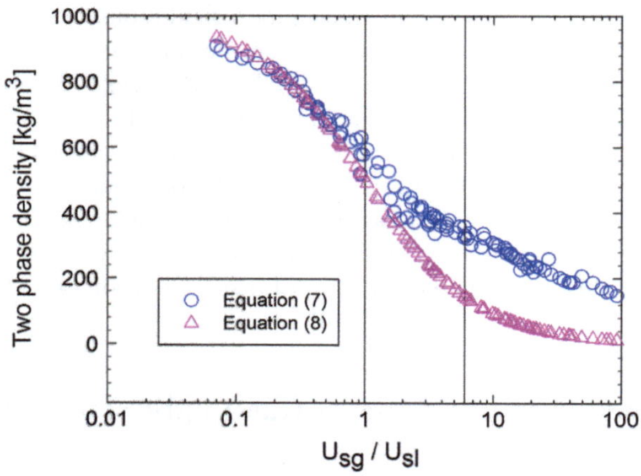

Fig. 4.14 Two phase mixture density as a function of void fraction and flow quality

in which it is assumed that the two phases flow with the same velocity i.e., it assumes no slip between the gas and liquid phase. However, in practise, with an exception of very small region of void fraction (bubbly flow region) there exists a significant slip between the two phases.

$$\rho_{tp} = \left(\frac{x}{\rho_g} + \frac{1-x}{\rho_l}\right)^{-1} \tag{4.8}$$

As shown in Fig. 4.14, the two phase density calculated using void fraction and quality, Eqs. (4.7) and (4.8) are in agreement for the small region of void fraction ($U_{sg}/U_{s1} \approx 1$) but shows a significant discrepancy for annular flow regime ($U_{sg}/U_{s1} > 1$). Thus it can be concluded that the use of homogeneous flow model to predict two phase mixture density may be acceptable only for the small region of void fraction where the slip between the two phases is negligible and in case of considerable slip between the two phases mixture density calculated by Eq. (4.7) should be preferred. The data plotted in Fig. 4.14 is for air water fluid combination measured at Two Phase Flow Lab, Oklahoma State University. Similar interaction between Eqs. (4.7) and (4.8) is expected for different fluid combinations.

The accelerational component of two phase pressure drop is due to the expansion of gas phase as the two phase mixture travels downstream and is trivial for non-boiling two phase flow and hence can be neglected. However, in case of boiling two phase flow; the pressure drop due to acceleration of the gas phase can offer a significant contribution to the total pressure drop depending upon the pipe diameter and pipe orientation. It should be noted that similar to hydrostatic component, the calculation of accelerational component of two phase pressure drop also requires correct estimation of void fraction at pipe inlet and exit flow conditions. The frictional component of two phase pressure drop is essentially due to the friction at pipe wall and gas-liquid interface and is the most complex and difficult to predict due to its dependency on several parameters such as, pipe orientation, surface roughness, pipe geometry, fluid properties, and flow patterns. The accelerational and frictional pressure drop components are expressed by Eqs. (4.9) and (4.10), respectively.

$$\left(\frac{dP}{dL}\right)_a = \frac{1}{L}\left\{\left[\frac{G_l^2}{\rho_l(1-\alpha)} + \frac{G_g^2}{\rho_g\alpha}\right]_{out} - \left[\frac{G_l^2}{\rho_l(1-\alpha)} + \frac{G_g^2}{\rho_g\alpha}\right]_{in}\right\} \tag{4.9}$$

$$\left(\frac{dP}{dL}\right)_{tp,f} = \frac{f_{tp}G^2}{2D\rho_{tp}} \tag{4.10}$$

4.6 Void Fraction and Hydrostatic Pressure Drop

As mentioned earlier, the direct influence of the void fraction on two phase hydrostatic pressure drop is through the mixture density and hence it is important to understand the sensitivity of two phase hydrostatic pressure drop to the void fraction. The two phase mixture density given by Eq. (4.7) is of weighted nature and is found to be sensitive to the large values of void fraction compared to the small values of void fraction. From Eq. (4.6), it is clear that the effect of error in void fraction on the calculation of hydrostatic pressure drop is maximum for vertical pipe orientation. As shown in Fig. 4.15, two phase hydrostatic pressure drop is more sensitive to the error in void fraction for the large values of void fraction in vertical pipe orientation. Whereas, the same error of ±10 % in low region of void fraction virtually doesn't influence the hydrostatic pressure drop. Since, the total pressure drop is of additive nature, it is desired to keep the error in each of these components as low as possible. In order to achieve this accuracy, it is required to have a correlation that will predict the void fraction with a minimum error for the large values of void fraction (annular flow). The correlation of [105] presented in the previous section satisfies this condition and is hence recommended for use for high values of void fraction.

It is also worthwhile to mention that the effect of error in void fraction on two phase hydrostatic pressure drop and hence on the overall accuracy in calculation of total two phase pressure drop may worsen in case of large pipe diameters. For small diameter pipes, compared to the hydrostatic pressure drop usually the frictional pressure drop has a greater share to the total two phase pressure drop and hence any error induced in hydrostatic pressure drop due to the error in void fraction is damped out. However, with increase in the pipe diameter, the share of hydrostatic pressure drop to the total two phase pressure drop becomes considerable and hence noticeably contributes to the total error. A better illustration of this scenario is reported by Ghajar and Bhagwat [41].

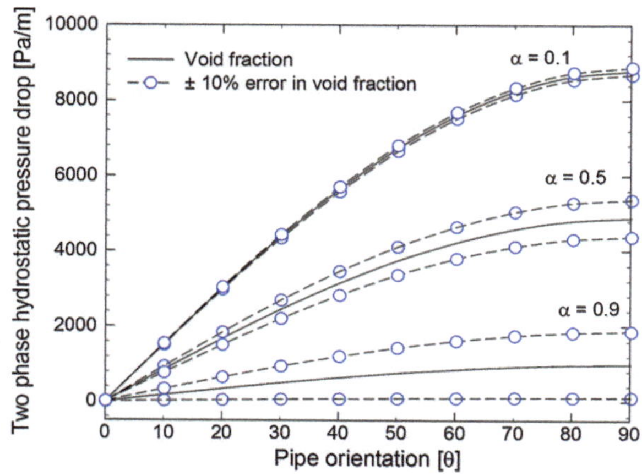

Fig. 4.15 Effect of void fraction error on two phase hydrostatic pressure drop (Eq. 4.6) for different pipe orientations

4.6.1 Effect of Flow Patterns and Pipe Diameter on Two Phase Frictional Pressure Drop

The frictional pressure drop in gas liquid two phase flow is essentially due to the friction of liquid or gas phase at the pipe wall and the friction at gas liquid interface. The relative magnitudes of the friction at pipe wall and at the gas liquid interface depend on the physical structure of individual flow patterns. For instance, the contribution of the interfacial friction to the two phase frictional pressure drop is significant when there is significant slippage between the two phases as in case of stratified and annular flow regimes. On the other hand, the gas liquid interfacial friction is trivial when the one phase is dispersed in another continuous phase and hence there is negligible slippage between the two phases as in case of bubbly flow regime. Consequently, the two phase frictional pressure drop in bubbly flow regime is mostly due to the friction of single phase liquid at the pipe wall and hence magnitude wise is only about 1–2 times higher than that of single phase flow. These higher magnitudes of two phase pressure drop in bubbly flow regime (small region of void fraction), are due to the turbulent eddies and the disturbances created in continuous liquid medium by the dispersed bubbles. For the other extreme of the two phase flow regime i.e., annular flow (separated flow), the magnitude of the interfacial friction is very high causing the two phase frictional pressure drop to be significantly higher (up to 1000 times depending upon the pipe diameter) than the corresponding single phase pressure drop measured at equivalent mass flow rates. The relationship between the two phase frictional pressure drop in terms of two phase frictional multiplier with the flow patterns in horizontal pipe orientation is evident from Fig. 4.16. The two phase frictional multiplier (Φ_l^2) is the ratio of the two phase frictional pressure drop to the pressure drop of single phase liquid or gas when assumed to flow through the pipe. The different forms of two phase frictional multiplier are discussed later in this work. It is observed that

Fig. 4.16 Two phase frictional pressure drop for different flow patterns in horizontal pipe orientation (air-water fluid combination)

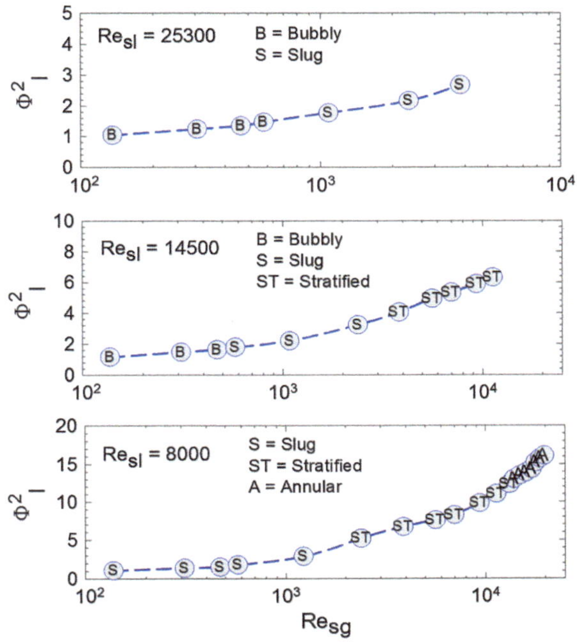

for the two phase pressure drop data measured in a 12.5 mm I.D. pipe using air-water fluid combination, the two phase frictional multiplier is close to unity for bubbly flow while is about 10–15 for annular flow regime. Similar relationship between the two phase frictional multiplier and the flow patterns exists for different pipe orientations and fluid combinations.

Similar to the single phase pressure drop, the frictional pressure drop in gas-liquid two phase flow is inversely proportional to the pipe diameter. However, as shown in the previous sections, the effect of pipe diameter on two phase frictional pressure drop at fixed pipe orientation is of different magnitude for different two phase flow regimes. It is found that the two phase frictional pressure drop in annular flow regime depends upon the pipe diameter to a great extent whereas it is relatively less sensitive to the pipe diameter in bubbly flow regime. As shown in Fig. 4.17, it is clear that for horizontal two phase flow at fixed mass flow rate of R134a, the two phase frictional pressure drop for different pipe diameters deviates significantly for large values of flow quality (annular flow regime) whereas, for low mass qualities, the pipe diameter has little effect on two phase frictional pressure drop. Similar conclusions were drawn from the work of [55] who compared the two phase frictional pressure drop of air-water in vertical upward flow for pipe diameters in a range of 10 mm < D < 50 mm. As mentioned earlier, the frictional pressure drop is due to the friction at pipe wall and friction at gas-liquid interface. The gas liquid interfacial area increases with increase in the pipe diameter and thus although the total two phase frictional pressure drop for larger

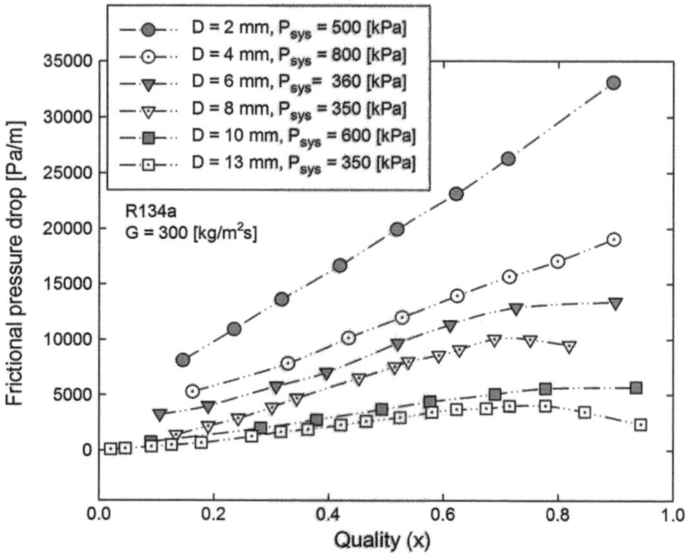

Fig. 4.17 Effect of pipe diameter on two phase frictional pressure drop in horizontal pipe orientation

pipe diameters is less than that compared to the smaller pipe diameters, the two phase frictional multiplier for large diameter pipes may be greater than that for the relatively smaller diameter pipes. Thus it is evident that any correlation developed to predict two phase frictional pressure drop must account for the pipe diameter effect in the high quality region or alternatively the annular flow regime.

4.6.2 Effect of Pipe Orientation on Two Phase Frictional Pressure Drop

In addition to the flow patterns and pipe diameter, the pipe orientation is also found to influence the two phase frictional pressure drop. For vertical two phase flow, the effect of pipe orientation on two phase frictional pressure drop is observed in vertical downward bubbly flow in form of coring phenomenon. This effect can be very well illustrated by presenting the measured two phase frictional pressure drop in form of two phase frictional multiplier [introduced later through Eq. (4.14)]. As shown in Fig. 4.18, for air-water vertical downward bubbly flow, the non-dimensional pressure drop in form of two phase frictional multiplier (Φ_l^2) is initially less than unity and then increases with increase in the gas superficial Reynolds number. This means that at the onset of bubbly flow the two phase

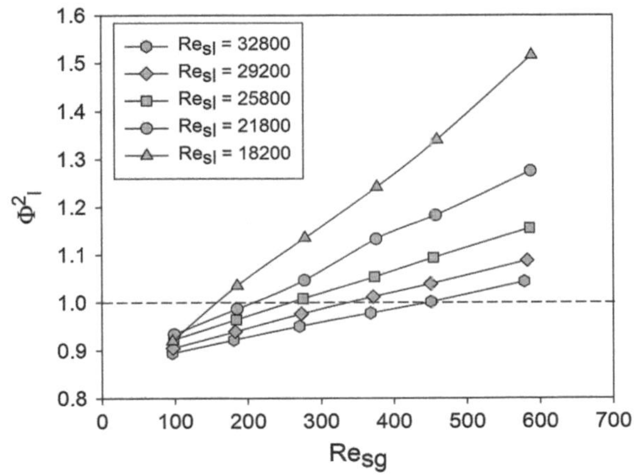

Fig. 4.18 Non-dimensional two phase frictional pressure drop in vertical downward bubbly flow (air-water data in 12.5 mm I.D. pipe)

frictional pressure drop is less than the single phase pressure drop calculated using Eq. (4.12). This trend in the data is essentially due to the coring phenomenon mentioned earlier in the section of flow patterns. With increase in both gas and liquid flow rates, the coring phenomenon gradually goes away and the two phase frictional pressure drop becomes higher than the single phase pressure drop yielding Φ_l^2 greater than unity. This phenomenon is not observed in the vertical upward and horizontal flow and hence the two phase frictional pressure drop increases with increase in the gas flow rate and is always greater than unity. Due to the lack of data in the literature, the trend of the two phase pressure drop in vertical downward flow and the existence of the coring phenomenon for different fluid combinations could not be verified.

Due to limited information on the flow patterns associated with the two phase pressure drop data, it is difficult to compare the effect of pipe orientation on two phase pressure drop for each flow pattern. However, the graphical representation of the variation of two phase frictional multiplier with respect to the Lockhart-Martinelli parameter [66] gives enough idea about the effect of pipe orientation on two phase frictional pressure drop. The Lockhart-Martinelli parameter is essentially the ratio of single phase liquid and gas frictional pressure drop as expressed by Eq. (4.11).

$$X = \sqrt{(dP/dL)_l \Big/ (dP/dL)_g} \tag{4.11}$$

The single phase pressure drop of liquid and gas phase is found using Eqs. (4.14) and (4.15). Thus from the structure of Eq. (4.11) it is evident that the low values of X represent the annular flow regime while the large values of X indicate existence of bubbly flow regime. The intermediate values of X are occupied by slug, stratified, froth and churn flow regimes. As shown in Fig. 4.19,

Fig. 4.19 Two phase frictional multiplier as a function of Lockhart-Martinelli parameter [66] for different pipe orientations (air-water data)

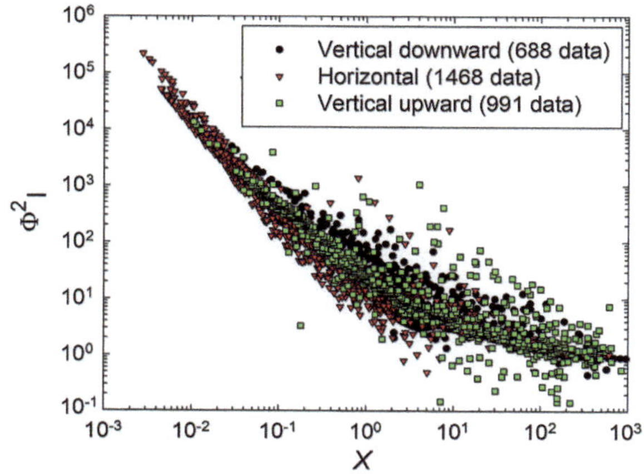

the large scatter for $X > 0.1$ indicates that the effect of pipe orientation is significant for non-annular flow patterns (buoyancy and gravity effect dominated flow). For $X < 0.1$ typically the annular flow regime (inertia effect dominated flow), the effect of pipe orientation on two phase frictional pressure drop gradually goes away. Some of the data for vertical upward pipe orientation that assumes the two phase frictional multiplier values less than unity are essentially due to the flow reversal phenomenon mostly observed for the churn flow regime. The flow reversal in churn flow regime is associated with the interfacial shear between the gas and liquid phase and the shear stress exerted by the liquid phase on pipe wall. For the sake of brevity the details of the flow reversal and its effect on two phase frictional pressure drop is not included in this chapter. Readers are advised to refer to the comprehensive work of [51, 68] for more details.

Due to the limited data available in the literature for different pipe diameters and orientations, it is difficult to extract the information about the combined effect of pipe orientation and the pipe diameter. However, based on the flow physics and the fact that there exist both buoyancy dominated and inertia dominated two phase flow regimes, it can be speculated that the effect of pipe orientation on two phase frictional pressure drop in buoyancy driven flows is more pronounced for large pipe diameters compared to the small diameter pipes. The two phase frictional pressure drop data shown in Fig. 4.19, is for air-water fluid combination and consists of a range of pipe diameters, 12.5 mm < D < 152 mm, 9.5 mm < D < 50 mm and 12.5 mm < D < 45.5 mm for horizontal, vertical upward and vertical downward flows, respectively. Thus the general trend of the data in Fig. 4.19 for small values of X gives enough evidence that the two phase frictional pressure drop in annular flow regime is relatively insensitive to the combined effect of pipe orientation and pipe diameter. Similar relationship between the two phase frictional multiplier and the Lockhart-Martinelli parameter is expected for other fluid combinations.

4.7 Review of Two Phase Frictional Pressure Drop Correlations

The prediction of two phase hydrostatic and accelerational pressure drop relies on the correct estimation of void fraction whereas the determination of frictional pressure drop in gas liquid two phase flow is much more complex. The two phase frictional pressure drop is determined either by finding a two phase friction factor (homogeneous flow model) or a two phase friction multiplier (separated flow model). These two ways of finding two phase frictional pressure drop are described below.

4.7.1 Homogeneous Flow Model Approach

The homogeneous flow model (HFM) assumes the two phases to be well mixed with each other and move with identical velocity (no interfacial slip). Thus in the homogeneous model the underlying idea is to represent the two phase mixture to behave as a pseudo single phase fluid that has fluid properties as that of two phase mixture. Thus HFM can be used with sufficient accuracy only when there is no rapid change in the flow variables and when there is negligible slip between the two phases.

The frictional pressure drop using homogeneous flow model is calculated with conventional equations used for single phase flow using pseudo mixture properties. For single phase flow, the frictional pressure drop is calculated using generalized equation of following form,

$$\left(\frac{dP}{dL}\right)_j = \frac{f_j G^2}{2D\rho_j} \tag{4.12}$$

Most of the homogeneous flow model based two phase frictional pressure drop correlations (two phase dynamic viscosity models listed in Table 4.11) recommend use of [14] correlation to determine friction factor.

$$f_j = 64/\mathrm{Re}_j \text{ for laminar flow } \left(Re_j < 2000\right)$$

$$f_j = 0.316/\mathrm{Re}_j^{0.25} \text{ for turbulent flow } \left(Re_j > 2000\right)$$

The subscript ‘j’ represents gas or liquid phase. This general form of the frictional pressure drop equation can be used to calculate two phase frictional pressure drop using two phase friction factor based on the two phase Reynolds number and the two phase mixture density as shown in Eq. (4.8). The two phase Reynolds number ($\mathrm{Re_m}$) is based on the two phase viscosity (μ_m) defined by Eq. (4.13).

$$\mathrm{Re}_m = \frac{GD}{\mu_m} \tag{4.13}$$

Table 4.11 Two phase dynamic viscosity models

Correlation	Expression for two phase dynamic viscosity
[1]	$\mu_m = \mu_l / (1-x) + x\sqrt{\rho_l/\rho_g}$
[9][a]	$\mu_m = \mu_l(1-\beta)(1+2.5\beta) + \mu_g\beta$
[24]	$\mu_m = x\mu_g + (1-x)\mu_l$
[28]	$\mu_m = \mu_l(1 + x(\rho_l/\rho_g - 1))$
[30]	$\mu_m = \rho_m(x(\mu_g/\rho_g) + (1-x)(\mu_l/\rho_l))$
[36]	$\mu_m = (1-\beta)\mu_l + \beta\mu_g + 2\sqrt{\beta(1-\beta)\mu_l\mu_g}$
[40]	$\mu_m = (\mu_l\rho_g/x\rho_l + (1-x)\rho_g)$
[64]	$\mu_m = \mu_l\mu_g/\mu_g + x^{1.4}(\mu_l - \mu_g)$
[71]	$\mu_m = (x/\mu_g + (1-x)/\mu_l)^{-1}$
[78]	$\mu_m = (\mu_l(1-\beta) + \mu_g\alpha/(1-\beta+\alpha))$
[4] (Model 1)[b]	$\mu_m = \mu_l \dfrac{2\mu_l + \mu_g - 2(\mu_l - \mu_g)x}{2\mu_l + \mu_g + (\mu_l - \mu_g)x}$
[4] (Model 2)[b]	$\mu_m = \mu_g \dfrac{2\mu_g + \mu_l - 2(\mu_g - \mu_l)(1-x)}{2\mu_g + \mu_l + (\mu_g - \mu_l)(1-x)}$
[4] (Model 3)[b]	Arithmetic mean of Model 1 and Model 2
[4] (Model 4)[b]	$\mu_m = \frac{1}{4}\left(\begin{array}{c}(3x-1)\mu_g + [3(1-x)-1]\mu_l \\ +\sqrt{[(3x-1)\mu_g + (3\{1-x\}-1)\mu_l]^2 + 8\mu_g\mu_l}\end{array}\right)$

[a] Uses [27] friction factor correlation
[b] Uses [23] friction factor correlation

Thus to predict the two phase frictional pressure drop correctly, an appropriate estimate of two phase Reynolds number and hence the two phase dynamic viscosity is obviously required. Literature reports several models to calculate the two phase dynamic viscosity as reported in Table 4.11.

4.7.2 Separated Flow Model Approach

Separated flow model accounts for the two phases flowing separately with different velocities and sharing a definite interface between them. Analytical solution to the separated flow model requires in all six equations; mass, momentum and energy conservation equations for each phase. Additional information such as velocity and temperature profile and other hydrodynamic parameters are required to solve these equations thus making it complex and difficult. An easy and a quick approach is to use empirical methods based on extensive experimental data.

The empirical model based on the concept of separated flow was first conceived by Lockhart and Martinelli [66] and since then several investigators have proposed different correlations by modifying the correlation of [66]. The separated flow model can be expressed in four different ways in terms of the two phase friction multiplier that accounts for the pressure drop due to the flow of single phase liquid or gas. The subscripts '*lo*' and '*go*' correspond to frictional pressure drop when the single phase liquid or gas flow rate is assumed to be equivalent to the entire two phase mixture flow rate (G). Whereas, the subscripts '*l*' and '*g*' indicate the frictional pressure drop when single phase liquid or gas is flowing at a rate of $G(1-x)$ and Gx, respectively. The understanding of these four approaches that can be used in separated flow model becomes more apparent from their mathematical definitions presented by Eqs. (4.14) to (4.17).

$$\left(\frac{dP}{dL}\right)_{tp,f} = \Phi_l^2\left(\frac{dP}{dL}\right)_l \text{ where } \left(\frac{dP}{dL}\right)_l = \frac{f_l G^2(1-x)^2}{2D\rho_l} \tag{4.14}$$

$$\left(\frac{dP}{dL}\right)_{tp,f} = \Phi_g^2\left(\frac{dP}{dL}\right)_g \text{ where } \left(\frac{dP}{dL}\right)_g = \frac{f_g G^2 x^2}{2D\rho_g} \tag{4.15}$$

$$\left(\frac{dP}{dL}\right)_{tp,f} = \Phi_{lo}^2\left(\frac{dP}{dL}\right)_{lo} \text{ where } \left(\frac{dP}{dL}\right)_{lo} = \frac{f_{lo} G^2}{2D\rho_l} \tag{4.16}$$

$$\left(\frac{dP}{dL}\right)_{tp,f} = \Phi_{go}^2\left(\frac{dP}{dL}\right)_{go} \text{ where } \left(\frac{dP}{dL}\right)_{go} = \frac{f_{go} G^2}{2D\rho_g} \tag{4.17}$$

Literature provides several correlations to determine the two phase frictional multipliers as shown in the above equations to predict two phase frictional pressure drop. The two phase frictional pressure drop correlations based on separated flow models are listed in Table 4.12. Most of these correlations are based on Eqs. (4.14) and (4.16) and require the single phase friction factor to be determined using [14] correlation. Any other single phase friction factor correlation recommended by certain correlation is also listed in Table 4.12.

4.8 Assessment of Two Phase Frictional Pressure Drop Correlations

The performance of the two phase frictional pressure drop correlations based on homogeneous and separated flow model considered in Tables 4.11 and 4.12 is assessed against a comprehensive data of air water and liquid refrigerant and its vapor consisting of 3147 and 1685 data points, respectively. The air water data consists of horizontal, vertical upward and downward pipe orientations while the refrigerant liquid vapor data consists of horizontal pipe orientation only. The experimental data used for the performance assessment of two phase frictional pressure drop correlations against air-water and refrigerant liquid-vapor is summarized in Tables 4.13.

Table 4.12 Two phase frictional pressure drop correlations based on separated flow model

Source	Correlation	Comments
[66]	$\Phi_l^2 = 1 + \frac{C}{X} + \frac{1}{X^2}$ where C = 5, 10 12 and 20 for laminar-laminar, laminar-turbulent, turbulent-laminar and turbulent-turbulent single phase flow conditions, respectively	Developed for data in horizontal smooth pipes
[6]	$\Phi_l^2 = \left(\left(\frac{1}{1-x}\right)\left(1 - \Gamma_{Bf}\left(1 - \frac{\rho_g}{\rho_l}\right)\right)^{0.43}\left(1 + x\left(\frac{\rho_g}{\rho_l} - 1\right)\right)\right)^{1.75}$ where $\Gamma_{Bf} = \left(\frac{0.71+2.35\left(\rho_g/\rho_l\right)}{1+\left(\frac{1-x}{x}\right)\left(\rho_g/\rho_l\right)}\right)$	Mathematical representation of the graphical solution to the frictional pressure drop given by Barcozy [7]
[18]	$\Phi_{lo}^2 = A_{Fr} + \frac{1.262x^{0.6978}}{We_{go}^{0.1458}}\left(\frac{\rho_l}{\rho_g}\right)^{0.3278}\left(\frac{\mu_g}{\mu_l}\right)^{-1.181}\left(1 - \frac{\mu_g}{\mu_l}\right)^{3.477}$ where $A_{Fr} = (1-x)^2 + x^2\frac{\rho_l}{\rho_g}\frac{f_{go}}{f_{lo}}$	Based on condensation of halogenated refrigerants inside a horizontal tube
[20]	$\Phi_{lo}^2 = 1 + (Y^2 - 1)\left(B_{Ch}x^{(1-0.5n)}\left(1 - x^{(1-0.5n)}\right)\right) + x^{(2-n)}$ where n = 0.25 and $Y = \sqrt{(dP/dL)_{go}/(dP/dL)_{lo}}$ and B_{Ch} is defined based on the combined criteria of mass flux and Y values as reported by Chisholm [20]	Accounts for the variation in two phase frictional pressure drop for different mass flux
[38, 39]	(For $\theta = 0^o$ and $+90^o$) $\Phi_{lo}^2 = A_{Fr} + \frac{3.24x^{0.78}(1-x)^{0.22}\left(\rho_l/\rho_g\right)^{0.91}\left(\mu_g/\mu_l\right)^{0.19}\left(1-\left(\mu_g/\mu_l\right)\right)^{0.7}}{Fr_{lo}^{0.045}We_{lo}^{0.035}}$ (For $\theta = -90^o$) $\Phi_{lo}^2 = A_{Fr} + \frac{5.7x^{0.7}(1-x)^{0.14}\left(\rho_l/\rho_g\right)^{0.85}\left(\mu_g/\mu_l\right)^{0.36}\left(1-\left(\mu_g/\mu_l\right)\right)^{0.2}}{Fr_{lo}^{0.09}We_{lo}^{0.007}}$ where A_{Fr} is same as that of [18] except that the single phase friction factors are calculated using $f_{jo} = 0.25\left[0.86859\ln\left(\frac{Re_{sj}}{1.964\ln Re_{sj} - 3.8215}\right)\right]^{-2}$	Based on experimental data bank of 25000 data points for air-water and refrigerants
[48]	$\Phi_{lo}^2 = 1 + dP_{Fr}\left[\frac{\left(\rho_l/\rho_g\right)}{\left(\mu_l/\mu_g\right)^{0.25}} - 1\right]$ where $dP_{Fr} = f_{lo}\left(x + 4\left(x^{1.8} - x^{10}f_{lo}^{0.5}\right)\right)$ and $f_{lo} = f(Fr)$ as given by Gronnerud [48]	

(continued)

Table 4.12 (continued)

Source	Correlation	Comments
[75]	$\Phi_{lo}^2 = Y^2x^3 + (1-x)^{0.33}(1 + 2x(Y^2 - 1))$, where Y is same as that of [20]	Extrapolation between all liquid and all gas flow
[54]	$\Phi_{lo}^2 = 30.78x^{1.323}(1-x)^{0.477}P_r^{-0.7232}$ where P_r is the reduced pressure of given refrigerant.	Applicable for turbulent-turbulent region and based on flow of refrigerants in horizontal tube
[73]	$\Phi_l^2 = 1 + \frac{C}{X_{tt}} + \frac{1}{X_{tt}^2}$ where $C = 21(1 - \exp(-0.319D))$ Pipe diameter is in [mm]	Based on air-water and refrigerant data. Accounts for pipe diameter effect
[96]	$\Phi_l^2 = 1 + C/X_{tt}^{1.19} + 1/X_{tt}^2$ where, $C = 1.79(\mathrm{Re}_g/\mathrm{Re}_l)^{0.4}((1/x) - 1)^{0.5}$	Based on more than 2000 data for refrigerants and air-water fluid combinations and D < 13 mm
[103]	$\Phi_l^2 = 1 + C/X_{tt} + 1/X_{tt}^2$ where, $C = 4.566 \times 10^{-6} X^{0.128} \mathrm{Re}_{lo}^{0.938} (\rho_g/\rho_l)^{2.15} (\mu_l/\mu_g)^{5.1}$	Based on refrigerant data in 6.5 mm horizontal pipe. Applicable for 50 < G < 700 kg/m^2 s
[92]	$\Phi_{lo}^2 = 1 + (\Gamma^2 - 1)x^{1.75}(1 + 0.952\Gamma X_{tt}^{0.142})$ where, $\Gamma = \sqrt{\rho_l/\rho_g}(\mu_g/\mu_l)^{0.125}$	Based on experimental data of R12, R22, R134a and R32/R125 in horizontal pipe of D < 10 mm
[62]	$\Phi_{lo}^2 = 1.7 + \left((6.3 + Fr_{lo})^{0.89} \Big/ 1.04 Fr_{lo}^{0.52} We_{lo}^{-0.011}\right)(1/X_{tt})^{0.42}$	
[63]	$\Phi_{lo}^2 = \frac{0.36(0.6 + AFr_{tp})^{1.3}}{Fr_{tp}^{0.51} We_{tp}^{-0.031}} X_{tt}^{0.15}$ where $Fr_{tp} = (G^2/gD\rho_m^2)$ and $We_{tp} = (G^2D/\sigma\rho_m)$ and ρ_m is from Eq. (4.8) A is variable defined for two different ranges of mixture mass flux	Based on data for R134a and R123 refrigerants in 10 mm I.D. horizontal pipe

(continued)

Table 4.12 (continued)

Source	Correlation	Comments
[107]	$\Phi_{lo}^2 = \left\{Y^2x^3 + (1-x)^{0.33}(1+2x(Y^2-1))\right\}\left[1+1.54(1-x)^{0.5}La\right]$ where La is the Laplace constant defined as $La = \left(\sqrt{\sigma/g(\rho_l-\rho_g)}/D\right)$ Y is same as that of (Chisholm [20] but the single phase friction factor is calculated using [34] correlation	Based on 2600 data of 15 refrigerants for $0.8 < D < 19$ mm in horizontal flow
[111]	$\Phi_{lo}^2 = 1 + 4.2(Y^2-1)\left\{\frac{B}{We}x^{0.875}(1-x)^{0.875} + x^{1.75}\right\}$, where Y is same as that of [20]	Modification of [20] correlation. Developed for boiling CO_2 two phase flow in a 7.5 mm horizontal pipe and claims to predict the data within ±20 % error bands
[99]	$\Phi_{lo}^2 = 1 + (4.3Y^2-1)La\left\{x(1-x)^{0.875} + x^{1.75}\right\}$, where Y is same as that of [20] and La is the Laplace number used by Xu and Fang [107]	Applicable for R12, R134a and R113 in smooth tubes with $P_{sys} < 860$ kPa
[100]	$\Phi_l^2 = 1 + \frac{C}{X} + \frac{1}{X^2}$ where $C = 1.279 \times 10^{-9}\Lambda^{-1.96}\Psi Re_{lo}^{0.4}$ for $We_{go} \le We_{lo}^{0.14}$ Else $C = 1.386 \times 10^{-4}\Lambda^{-0.65}\Psi^{0.2}Re_{lo}^{0.52}$ where $\Psi = \frac{\mu_l}{\sigma}1.2\left(\frac{G_g}{\rho_g} + \frac{G_l}{\rho_l}\right)$ and $\Lambda = \mu_l^2/\rho_l\sigma D^2$	
[88]	$\Phi_{lo}^2 = 0.38Re_{lo}^{0.1}\left(1 + \frac{x}{1-x}\frac{U_g}{U_l}\right)^{0.95}$	
[69]	$\Phi_{lo}^2 = \Phi_l^2(1-x)^{1.75}$ where Φ_l^2 is obtained from [66]	Developed for boiling two phase flow of steam-water mixtures in horizontal pipes
[112]	$\Phi_l^2 = 1 + C/X_{tt}^{1.19} + 1/X_{tt}^2$ where $C = 21(1-\exp(-C_1/La))$ and $C_1 = 0.142$ and 0.674 for adiabatic liquid vapor and liquid and gas flow, respectively. La is the Laplace constant used by Xu and Fang [107]	Based on air-water and refrigerant data in horizontal and vertical upward mini and micro channels

Table 4.13 Summary of experimental data[a] used for performance assessment of two phase frictional pressure drop correlations

Parameter	Range
Air-water (1468 *data points*) ($\theta = 0^o$)	
Diameter (mm)	12.5–152
Pressure (kPa)	108–680
Mass flux (kg/m^2 s)	1.2–2620
Quality	0.0001–0.99
Void fraction	0.05–0.99
Air-water (991 *data points*) ($\theta = 90^o$)	
Diameter (mm)	9.5–50
Pressure (kPa)	102–575
Mass flux (kg/m^2 s)	1.1–3260
Quality	0.0001–0.93
Void fraction	0.03–0.99
Air-water (688 *data points*) ($\theta = -90^o$)	
Diameter (mm)	12.5–45.5
Pressure (kPa)	102–280
Mass flux (kg/m^2 s)	25–2450
Quality	0.0005–0.79
Void fraction	0.02–0.98
Refrigerant (1685 *data points*) ($\theta = 0^o$)	
Diameter (mm)	1.4–13.8
Pressure (kPa)	150–2650
Mass flux (kg/m^2 s)	25–1082
Quality	0.01–0.97
Void fraction	0.26–0.98
Fluids	CO_2, R12, R123, R134a, R22, R404a, R407C, R401A, R507

[a] Sources of the experimental data are not reported due to space constraints

The performance of frictional pressure drop correlations considered in this chapter is assessed based on the percentage of data points predicted within ±30 % error bands. The accuracy criteria of ±30 % is based on the overall performance of the correlations and is consistent with the assessment done by other investigators such as [84, 85, 96, 103 (used ±50 %)] and (Xu et al. [108]). The two phase pressure drop data for air water is divided into different ranges of the pipe diameter for each pipe orientation. The assessment of the frictional pressure drop correlations based on this classification gives an impression about their scope to account for the effect of pipe diameter on two phase frictional pressure drop. As shown in Table 4.14, the top performing correlations are short listed for the two phase frictional pressure drop data divided into four categories of the pipe diameter range. For the horizontal two phase flow and smallest pipe diameter of D = 12.5 mm [75] give best accuracy by predicting about 98 % of data within

Table 4.14 Top performing frictional pressure drop correlations for air-water two phase flow in horizontal pipe orientation

Diameter [mm]	Correlation	% of data within ±30 % error bands
D = 12.5	[75]	97.9
	[63]	95.4
	[4] (Model 2)	97.1
12.5 < D < 25	[73]	73.2
	[96]	73.2
	[9]	65.4
25 < D < 50	[73]	79.3
	[30]	78.5
	[9]	76.4
90 < D < 150	[73]	63.1
	[30]	66.5
	[9]	67.5

Table 4.15 Top performing frictional pressure drop correlations for air water two phase flow in vertical upward pipe orientation

Diameter [mm]	Correlation	% of data within ±30 % error bands
9 < D < 15	[69]	85.4
	[75]	82.3
	[4] (Model 2)	82.1
20 < D < 40	[73]	73.8
	[96]	73.8
	[9]	70.3
40 < D < 50	[38]	70.1
	[75]	66.2
	[4] (Model 3)	68.1

±30 % error bands. For the next three ranges of pipe diameter, the correlations of [9, 73] give consistent performance. The correlation of [73] is based on the concept of separated flow model while the correlation of [9] is essentially to predict two phase dynamic viscosity and hence is based on the concept of homogeneous two phase flow model. Overall it is observed that the performance of the both types of two phase frictional pressure drop correlations deteriorate with increase in the pipe diameter. A more detailed look at the performance of these correlations showed that the accuracy of these correlations decrease in particular for the annular flow regime. This is probably because, these correlations do not have enough versatality to account for the increase in the interfacial friciton in the annular regime with increase in the pipe diameter.

In case of vertical upward flow, for the lowest range of pipe diameter [69] give best performance by predicting more than 85 % of data within ±30 % error bands as shown in Table 4.15. For the intermediate range of pipe diameter i.e., 20 mm < D < 40 mm the correlation of [73] is one of the best correlations that predicts about 74 % of data within ±30 % error bands. Similar to the horizontal

Table 4.16 Top performing frictional pressure drop correlations for air water two phase flow in vertical downward pipe orientation

Diameter [mm]	Correlation	% of data within ±30 % error bands
10 < D < 20	[18]	76.8
	[20]	73.2
	[4] (Model 2)	71.4
20 < D < 35	[99]	49.8
	[112]	40.2
	[18]	44.2
40 < D < 50	[18]	48.3
	[20]	46.6
	[111]	43.3

pipe orientation the accuracy of the two phase frictional pressure drop correlations is found to decrease with increase in the pipe diameter. Table 4.16 reports the top performing correlations for three different ranges of the pipe diameter. In comparison to the horizontal and vertical upward flows, all the correlations predict the two phase frictional pressure drop data with a lower accuracy for vertical downward flow. The correlation of [18] is found to be one of the top performing correlations for all three ranges of the pipe diameter. The large inaccuracies for the pipe diameter range of 20 mm < D < 40 mm and 40 mm < D < 50 mm are probably due to the fact that most of the data for these pipe diameters belong to the annular flow regime and as mentioned earlier the two phase pressure drop correlations are found not to perform well in this flow regime. Overall, independent of the pipe orientation it is observed that the two phase pressure drop correlations do not predict the data correctly for large pipe diameters and especially in annular flow regimes.

The top performing correlations for two phase flow of nine different refrigerants in horizontal pipe orientation are shortlisted in Table 4.17. It is found that the correlations of [28, 107] are among the top performing correlations for majority of the refrigerants. The best performance is given by the correlations for R404A and R407 C with 26 and 42 data points, respectively. More data is required for these refrigerants to verify and confirm the accuracy of short listed top performing correlations. In comparison to other refrigerants, all the correlations considered in this chapter are found to give less accuracy for refrigerant R134a with 514 data points. Overall it appears that the correlations of [28, 107] have potential to predict the two phase frictional pressure drop correctly over a wide range of refrigerants and hence must be considered for any further modification to improve their accuracy. Although the correlation of [107] is not among the top performing correlations for R123, R407C, and R507 it predicts 61, 81 and 67 % of data within ±30 % error bands. A more comprehensive analysis is required to determine the accuracies of these correlations for different flow patterns or alternatively different ranges of the void fraction for both air-water and refrigerant data.

Table 4.17 Top performing frictional pressure drop correlations for refrigerant liquid vapor two phase flow in horizontal pipe orientation

Refrigerant	Correlation	% of data within ±30 % error bands	Standard deviation (%)
CO_2	[28]	87	31
(287 data points)	[107]	85	32
	[38]	76	52
R12	[28]	85	29
(161 data points)	[107]	78	26
	[38]	71	36
R123	[28]	71	34
(42 data points)	[103]	74	32
	[18]	71	19
R134a	[28]	68	47
(514 data points)	[107]	67	47
	[38]	62	52
R22	[18]	72	46
(326 data points)	[28]	75	40
	[107]	69	44
R404A	[64]	96	24
(26 data points)	[9]	88	22
	[4] (Model 2)	92	25
R407C	[4] (Model 1)	95	20
(42 data points)	[92]	93	18
	[73]	95	19
R401A	[28]	75	37
(229 data points)	[107]	68	39
	[92]	68	39
R507	[18]	79	27
(58 data points)	[38]	74	27
	[28]	69	31

4.9 Application of the Recommended Void Fraction and Two Phase Frictional Pressure Drop Correlations

The objective of this section is to include some of the fundamental equations covered in this work and guide readers to select the appropriate and correct methods to solve two phase flow problems. The two problems considered here are for the air-water and refrigerant two phase flow that requires determination of void fraction and two phase pressure drop.

(1) *Consider a two component two phase flow of air-water flowing upwards in a* 2.54 cm *I.D. pipe* 15° *inclined from horizontal. The total mass flux of the mixture is* 600 kg/m^2 s *at system pressure of 7 bar and temperature of* 25 °C. *It is assumed that the mass flow quality of the two phase mixture remains constant at 0.095 throughout the pipe length. Calculate (a) void fraction (b) slip ratio and (c) the two phase mixture density. Verify if the use of homogeneous model approach to calculate the two phase mixture density is appropriate for this two phase flow situation.*

Solution: For the system pressure of 7 bar and 25 °C, the physical properties of air and water are as follows.

$$\rho_l = 997.3(\text{kg/m}^3),\ \rho_g = 8.196(\text{kg/m}^3),\ \sigma = 0.0719(\text{N/m})$$

The mass flux of each liquid (water) and gas (air) phase is calculated as follows,

$$G_l = G(1-x) = 600\left(\text{kg/m}^2\text{s}\right) \times (1-0.095) = 543\left(\text{kg/m}^2\text{s}\right)$$

$$G_g = Gx = 600\left(\text{kg/m}^2\text{s}\right) \times 0.095 = 57\left(\text{kg/m}^2\text{s}\right)$$

From the liquid and gas mass flux, the superficial velocity for each phase is calculated as,

$$U_{sl} = \frac{G_l}{\rho_l} = \frac{543\left(\text{kg/m}^2\text{s}\right)}{997.3\left(\text{kg/m}^3\right)} = 0.544(\text{m/s}),\ U_{sg} = \frac{G_g}{\rho_{gl}} = \frac{57\left(\text{kg/m}^2\text{s}\right)}{8.196\left(\text{kg/m}^3\right)}$$
$$= 6.95(\text{m/s}) \text{ and}$$

$$U_m = U_{sl} + U_{sg} = 0.544(\text{m/s}) + 6.95(\text{m/s}) = 7.494(\text{m/s})$$

(a) The selection of the appropriate void fraction correlation is based on the criteria of gas volumetric fraction (β) and void fraction (α) as recommended in Table 4.10.

The gas volumetric flow fraction is $\beta = \frac{U_{sg}}{U_{sl}+U_{sg}} = \frac{6.95}{6.95+0.544} = 0.927$

From Table 4.10, for 15° upward inclined pipe orientation and $\beta = 0.927$, correlations of [12, 105] are recommended for two different ranges of void fraction. However, as mentioned earlier, in case of β shared by two different ranges of void fraction, it is recommended to calculate the void fraction using both correlations and if the void fraction predicted by Woldesemayat and Ghajar [105] is greater than 0.75 then use the larger of two values for calculation purposes. Let α_1 and α_2 be the void fractions predicted by the [12, 105] correlations, respectively. Then the final value of void fraction to be considered for the calculation purposes is the maximum of two values i.e., $\alpha = \max(\alpha_1, \alpha_2)$.

Since the correlation of [12] is implicit in nature (see Table 4.2), the void fraction is calculated on an iterative basis.

The iterative calculation yields, $C_o = 1.073$ and $U_{gm} = 0.64$ (m/s)

Thus the void fraction from Eq. (4.4) is $\alpha_1 = \frac{U_{sg}}{C_o U_m + U_{gm}} = \frac{6.95}{(1.073 \times 7.49) + 0.64} = 0.799$.

Whereas, the calculation of void fraction using the correlation of [105] given in Table 4.2 is as follows,

$$C_o = \frac{U_{sg}}{U_m}\left(1 + \left(\frac{U_{sl}}{U_{sg}}\right)^{b_1}\right) = \frac{6.95}{7.49}\left(1 + \left(\frac{0.544}{6.95}\right)^{0.618}\right) \text{ where } b_1 = \left(\frac{\rho_g}{\rho_l}\right)^{0.1} = \left(\frac{8.196}{997.3}\right)^{0.1} = 0.618$$

$$C_o = 1.119$$

$$U_{gm} = 2.9\left[gD\sigma(1 + \cos\theta)\Delta\rho/\rho_l^2\right]^{0.25}(1.22 + 1.22\sin\theta)^{b_2} \text{ where } b_2 = \frac{P_{atn}}{P_{sys}} = \frac{1.013}{7} = 0.1447$$

$$U_{gm} = 2.9\text{m}^{-0.25}\left(9.81(\text{m/s}^2)0.0254(\text{m}) \times 0.0719(\text{N/m}) \times (1 + \cos(15)) \times \frac{(997.3 - 8.196)}{(997.3)^2}\right)^{0.25} \times (1.22 + 1.22\sin(15))^{0.1447}$$

$$U_{gm} = 0.237(\text{m/s})$$

Thus the void fraction is calculated using Eq. (4.4),

$$\alpha_2 = \frac{U_{sg}}{C_o U_m + U_{gm}} = \frac{6.95}{(1.119 \times 7.49) + 0.237} = 0.806$$

The predictions of [12, 105] are in very good agreement with the predictions of other two top performing correlatons for $\alpha > 0.75$ and $\beta > 0.9$. The other two top performing correlation for this range from Table 4.7 are that of [25, 91] that predict void fraction of 0.81 and 0.779, respectively. Thus as mentioned earlier, since β is shared by two different ranges of void fraction and since the predicted value of void fraction is greater than 0.75, we will select the larger of the two values.

Thus the void fraction for given two phase flow condition is, $\alpha = \alpha_2 = 0.806$.

(b) The slip ratio is the ratio of the actual velocity of the gas phase to the actual velocity of liquid phase expressed by Eq. (4.2),

$$S = \frac{U_g}{U_l} = \frac{(U_{sg}/\alpha)}{(U_{sl}/(1 - \alpha))} = \frac{(6.95/0.806)}{(0.544/(1 - 0.806))} = \frac{8.62}{2.804} = 3.07$$

(c) The two phase mixture density is calculated using Eq. (4.7) as follows,

$$\rho_{tp} = \rho_l(1 - \alpha) + \rho_g\alpha = 997.3(\text{kg/m}^3) \times (1 - 0.806) + 8.196(\text{kg/m}^3) \times (0.806) = 200.1(\text{kg/m}^3)$$

Another approach to calculate two phase mixture density is using Eq. (4.8) i.e., assuming homogeneous two phase flow (no slip between the two phases) as shown below.

$$\rho_{tp} = \left(\frac{x}{\rho_g} + \frac{1 - x}{\rho_l}\right)^{-1} = \left(\frac{0.095}{8.196} + \frac{1 - 0.095}{997.3}\right)^{-1} = 80.01(\text{kg/m}^3)$$

It is clear that the two phase mixture density calculated using homogeneous flow model approach is about one third of the actual density calculated using void fraction. It is shown earlier and illustrated in Fig. 4.14 that the use of homogeneous flow model to calculate two phase mixture density is appropriate when the slip ratio is close to unity. However, in this two phase flow situation, the slip ratio is 3 and hence the use of homogeneous flow model i.e., Eq. (4.8) to calculate the two phase mixture density is inappropriate.

(2) *Consider a boiling two phase flow of R134a through a 6 mm I.D. horizontal copper tube at a mass flux of* 800 kg/m^2 s *and a system pressure of 1500* kPa. *The two phase refrigerant enters the 1 m long tube at* $x = 0.3$ *and exits at* $x = 0.7$ *and assume that the quality changes linearly with the pipe length. For simplicity calculate the phase velocities, average void fraction and the frictional pressure drop at a mean quality of 0.5. Using the appropriate correlations of void fraction and frictional pressure drop, calculate (a) the average value of void fraction, two phase mixture density and hydrostatic pressure drop (b) accelerational pressure drop across the pipe length accounting for the change in two phase quality at inlet and exit (c) frictional and total two phase pressure drop and (d) what are the contributions of different components of two phase pressure drop to the total two phase pressure drop?*

Solution: For refrigerant R134a at 1500 kPa, the thermo physical properties of liquid and vapor phase are as follows,

Saturation temperature = 55.5 °C, Surface tension (σ) = 0.00427 N/m
Liquid density (ρ_l) = 1078 (kg/m^3), Vapor density (ρ_g) = 76.95 (kg/m^3),
Liquid viscosity (μ_l) = 0.0001746 (Pa.s), Vapor viscosity (μ_g) = 0.0000138 (Pa.s).

(a) Based on the mean quality of x = 0.5 and the total mass flux of 800 kg/m^{2}s, the liquid and vapor phase superficial velocities are calculated as follows,

$$U_{sl} = \frac{G(1-x)}{\rho_l} = \frac{800(\text{kg/m}^2\text{s}) \times (1-0.5)}{1078(\text{kg/m}^3)} = 0.371(\text{m/s})$$

$$U_{sg} = \frac{Gx}{\rho_g} = \frac{800(\text{kg/m}^2\text{s}) \times 0.5}{76.95(\text{kg/m}^3)} = 5.198(\text{m/s}) \text{ and } U_m = U_{sl} + U_{sg} = 5.57(\text{m/s})$$

As recommend in Table 4.10, for horizontal two phase flow of refrigerants, the correlation of [105] can be used for the entire range of void fraction and gas volumetric flow fraction.

The distribution parameter and drift velocity required in this correlation as given in Table 4.2 is calculated as shown below,

$$C_o = \frac{U_{sg}}{U_m}\left(1 + \left(\frac{U_{sl}}{U_{sg}}\right)^{b_1}\right) = \frac{5.198}{5.57}\left(1 + \left(\frac{0.371}{5.198}\right)^{0.768}\right) \text{ where } b_1 = \left(\frac{\rho_g}{\rho_l}\right)^{0.1} = \left(\frac{76.95}{1078}\right)^{0.1} = 0.768$$

$$C_o = 1.056$$

$$U_{gm} = 2.9\left[gD\sigma(1+\cos\theta)\Delta\rho/\rho_l^2\right]^{0.25}(1.22 + 1.22\sin\theta)^{b_2} \text{ where } b_2 = \frac{P_{atm}}{P_{sys}} = \frac{101.32(kPa)}{1500(kPa)} = 0.0675$$

$$U_{gm} = 2.9\text{m}^{-0.25}\left(9.81(\text{m/s}^2)0.006(\text{m}) \times 0.00427(\text{N/m}) \times (1+\cos(0)) \times \frac{(1078 - 76.95)}{(1078)^2}\right)^{0.25}$$
$$\times (1.22 + 1.22\sin(0))^{0.0675}$$

$$U_{gm} = 0.0754(\text{m/s})$$

Thus the void fraction is calculated using Eq. (4.4) as shown below,

$$\alpha = \frac{U_{sg}}{C_o U_m + U_{gm}} = \frac{5.198}{(1.056 \times 5.57) + 0.0754} = 0.872$$

Based on the void fraction, the two phase mixture density is calculated using Eq. (4.7).

$$\rho_{tp} = \rho_l(1-\alpha) + \rho_g\alpha = 1078(\text{kg/m}^3) \times (1 - 0.872) + 76.95(\text{kg/m}^3) \times (0.872)$$
$$= 205.1(\text{kg/m}^3)$$

For horizontal pipe orientation, the hydrostatic pressure drop is zero. i.e., $\left(\frac{dP}{dL}\right)_h = 0$. Otherwise for any other orientation it is calculated using Eq. (4.6).

(b) Given that the flow quality changes from 0.3 to 0.7 from pipe inlet to exit, the acceleration component of two phase pressure drop is calculated using Eq. (4.9).

$$\left(\frac{dP}{dL}\right)_a = \frac{1}{L}\left\{\left[\frac{G_l^2}{\rho_l(1-\alpha)} + \frac{G_g^2}{\rho_g\alpha}\right]_{out} - \left[\frac{G_l^2}{\rho_l(1-\alpha)} + \frac{G_g^2}{\rho_g\alpha}\right]_{in}\right\}$$

For the inlet and exit quality of 0.3 and 0.7, the liquid and vapor mass flux are calculated as follows,

At pipe inlet for $x = 0.3$,

$$G_l = G(1-x) = 800(\text{kg/m}^2\text{s}) \times (1 - 0.3) = 560(\text{kg/m}^2 s),$$
$$G_g = Gx = 800(\text{kg/m}^2 s) \times 0.3 = 240(\text{kg/m}^2\text{s})$$

Similarly at the pipe outlet for $x = 0.7$,

$$G_l = G(1-x) = 800(\text{kg/m}^2\text{s}) \times (1 - 0.7) = 240(\text{kg/m}^2\text{s}),$$
$$G_g = Gx = 800(\text{kg/m}^2\text{s}) \times 0.7 = 560(\text{kg/m}^2\text{s})$$

Based on the inlet and outlet mass flux, the superficial velocity of each phase is calculated as

At pipe inlet,

$$U_{sl} = \frac{G_l}{\rho_l} = \frac{560(\mathrm{kg/m^2 s})}{1078(\mathrm{kg/m^3})} = 0.519(\mathrm{m/s}),$$

$$U_{sg} = \frac{G_g}{\rho_g} = \frac{240(\mathrm{kg/m^2 s})}{76.95(\mathrm{kg/m^3})} = 3.118(\mathrm{m/s}),\ U_m = 3.637(\mathrm{m/s})$$

At pipe outlet,

$$U_{sl} = \frac{G_l}{\rho_l} = \frac{240(\mathrm{kg/m^2 s})}{1078(\mathrm{kg/m^3})} = 0.223(\mathrm{m/s}),$$

$$U_{sg} = \frac{G_g}{\rho_g} = \frac{560(\mathrm{kg/m^2 s})}{76.95(\mathrm{kg/m^3})} = 7.277(\mathrm{m/s}),\ U_m = 7.5(\mathrm{m/s})$$

Based on the phase superficial velocities, the distribution parameter using [105] at pipe inlet and outlet are found to be $C_o = 1.073$ and $C_o = 1.037$, respectively. The drift velocity of [105] correlation is independent of the phase velocities and hence remains unchanged i.e., $U_{gm} = 0.0754$ (m/s) at pipe inlet and outlet.

The void fraction based on the phase velocities and distribution parameter calculated above gives, $\alpha_{in} = 0.783$ and $\alpha_{out} = 0.926$.

The acceleration pressure drop is,

$$\left(\frac{dP}{dL}\right)_a = \left\{\left[\frac{240^2}{1078(1-0.926)} + \frac{560^2}{76.95 \times 0.926}\right]_{out} - \left[\frac{560^2}{1078(1-0.783)} + \frac{240^2}{76.95 \times 0.783}\right]_{in}\right\} = 2826.5\left(\frac{\mathrm{Pa}}{\mathrm{m}}\right)$$

(c) From Table 4.17, the recommended correlations for two phase flow of R134a in a horizontal pipe are those of [28, 38, 107]. The [28] correlation is based on the concept of homogeneous flow model while the correlations of [38, 107] are based on the separated flow model concept. We will use the correlation of [107] for this problem.

From Table 4.12, the two phase frictional multiplier for [107] is determined from the following equation,

$$\Phi_{lo}^2 = \left\{Y^2x^3 + (1-x)^{0.33}\left(1 + 2x(Y^2-1)\right)\right\}\left[1 + 1.54(1-x)^{0.5}La\right]$$

$$\text{where, } Y = \sqrt{\frac{(dP/dL)_{go}}{(dP/dL)_{lo}}} \text{ and } La = \sqrt{\left(\frac{\sigma}{g(\rho_l - \rho_g)} \Big/ D\right)}$$

It should be noted that the [107] recommend use of [34] correlation for determination of single phase friction factor f_{go} and f_{lo}.

The [34] correlation for single phase turbulent flow is given as,

$$f_{lo} = 0.25\left(\log\left(\frac{150.39}{\mathrm{Re}_{lo}^{0.98865}} - \frac{152.66}{\mathrm{Re}_{lo}}\right)\right)^{-2} \text{ and} f_{go} = 0.25\left(\log\left(\frac{150.39}{\mathrm{Re}_{go}^{0.98865}} - \frac{152.66}{\mathrm{Re}_{go}}\right)\right)^{-2}$$

The Reynolds number required in the above equation is defined for liquid and gas phases as shown in following equations,

$$\mathrm{Re}_{lo} = \frac{GD}{\mu_l} = \frac{800(\mathrm{kg/m^2s}) \times 0.006(m)}{0.0001746} = 27491 \text{ and}$$
$$\mathrm{Re}_{go} = \frac{GD}{\mu_g} = \frac{800(\mathrm{kg/m^2s}) \times 0.006(m)}{0.0000138} = 347826$$

Thus from the friction factor equation of [34] we get,

$$f_{lo} = 0.02398 \text{ and} f_{go} = 0.01407.$$

The single phase pressure drop for liquid and gas phases are calculated from Eqs. (4.16) and (4.17),

$$\left(\frac{dP}{dL}\right)_{lo} = \frac{f_{lo}G^2}{2D\rho_l} = \frac{0.02398 \times 800^2(\mathrm{kg/m^2s})^2}{2 \times 0.006(\mathrm{m}) \times 1078(\mathrm{kg/m^3})} = 1186\left(\frac{\mathrm{Pa}}{\mathrm{m}}\right) \text{ and}$$
$$\left(\frac{dP}{dL}\right)_{go} = \frac{f_{go}G^2}{2D\rho_g} = \frac{0.01407 \times 800^2(\mathrm{kg/m^2s})^2}{2 \times 0.006(\mathrm{m}) \times 76.95(\mathrm{kg/m^3})} = 9752\left(\frac{\mathrm{Pa}}{\mathrm{m}}\right)$$

Thus, $Y = \sqrt{\frac{9752(\mathrm{Pa/m})}{1186(\mathrm{Pa/m})}} = 2.867$ and the Laplace number is calculated to be, La = 0.1099.

Thus the two phase frictional multiplier is found to be,

$$\Phi_{lo}^2 = \left\{2.867^2 \times 0.5^3 + (1 - 0.5)^{0.33}(1 + 2 \times 0.5(2.867^2 - 1))\right\}\left[1 + 1.54(1 - 0.5)^{0.5}0.1099\right] = 8.455$$

Now, the two phase frictional pressure drop is determined using Eq. (4.16) as,

$$\left(\frac{dP}{dL}\right)_{tp,f} = \Phi_{lo}^2\left(\frac{dP}{dL}\right)_{lo} = 8.455 \times 1186\left(\frac{\mathrm{Pa}}{\mathrm{m}}\right) = 10028\left(\frac{\mathrm{Pa}}{\mathrm{m}}\right)$$

The total two phase pressure drop is the sum of hydrostatic, accelerational and frictional two phase pressure drops expressed by Eq. (4.5) as,

$$\left(\frac{dP}{dL}\right)_{t,tp} = \left(\frac{dP}{dL}\right)_h + \left(\frac{dP}{dL}\right)_f + \left(\frac{dP}{dL}\right)_a = 0 + 10028 + 2826.5 = 12854.5\left(\frac{\mathrm{Pa}}{\mathrm{m}}\right)$$

(c) It is found that the accelerational pressure drop contributes to about 22 % of the total two phase pressure drop while the remaining 78 % contribution is due to the frictional two phase pressure drop. It should be noted that in case of adiabatic two phase flow, the quality and void fraction at pipe inlet and exit remain practically unchanged and hence the accelerational two phase pressure drop may be neglected.

4.10 Concluding Remarks

The purpose of this work is fulfilled by a brief and insightful discussion about the two phase flow patterns, void fraction and pressure drop. The two phase flow patterns are elucidated with the help of still photographs and their transition from one flow pattern to another is explained with the help of flow pattern maps. The void fraction and two phase pressure drop data used for both air-water and refrigerants are one of the most comprehensive data ever used for the performance assessment of void fraction and two phase pressure drop correlations. This data bank has been of great help not only for the performance assessment of the correlations but has also contributed to better understanding of the two phase flow phenomenon through a parametric analysis of the two phase void fraction and pressure drop with respect to the variables such as, flow patterns, pipe diameter and pipe orientation.

A comprehensive scrutiny of the void fraction and two phase pressure drop correlations for several two phase flow conditions has resulted in recommendations of the top performing correlations for different cases. Based on the recommendations of Table 4.10, only two correlations of [12, 105], can be used to predict the void fraction with fair accuracy in both one component and two component two phase flow over a range of pipe diameters and pipe orientations. However, more experimental data is required in the low region of void fraction typically $\alpha < 0.5$ in both upward and downward inclined pipe orientations to validate and further improve their performance in these orientations. It is observed that, the correlation of [107] has good potential to predict the two phase frictional pressure drop and may be considered for further analysis and modification to account for the two phase flow of different refrigerants. This work is concluded by providing two sample problems that deal with air-water and refrigerant two phase flow under different operating conditions. The objective of this exercise is achieved by explaining the selection process of the recommended correlations for void fraction and two phase pressure drop in addition to highlighting some of the fundamental concepts.

Acknowledgements The authors are thankful to Dr. John Thome (EPFL, Switzerland), Dr. Josua Meyer (University of Pretoria, South Africa), Dr. Somchai Wongwises (KMUTT, Thailand) and Dr. Neima Brauner (Tel Aviv University, Israel) for sharing void fraction and pressure drop data.

References

1. W.W. Akers, A. Deans, O.K. Crossee, Condensing heat transfer within horizontal tubes. Chem. Eng. Prog. **55**, 171–176 (1959)
2. S. Al-lababidi, A. Addali, H. Yeung, D. Mba, F. Khan, Gas void fraction measurement in two phase gas liquid slug flow using acoustic emission technology. Trans. ASME **131**, 501–507 (2009)
3. P. Alia, L. Cravarolo, A. Hassid, E. Pedrocchi, Liquid volume fraction in adiabatic two-phase vertical upflow in round conduits C.I.S.E, 1965
4. M.M. Awad, Y.S. Muzychka, Effective property models for homogeneous two-phase flows. Exp. Thermal Fluid Sci. **33**, 106–113 (2008)
5. S. Badie, C.P. Hale, G.F. Hewitt, Pressure gradient and holdup in horizontal two phase gas-liquid flows with low liquid loading. Int. J. Multiph. Flow **26**, 1525–1543 (2000)
6. S.G. Bankoff, A variable density single fluid model for two phase flow with particular reference to steam-water flow. J. Heat Transfer **11**, 165–172 (1960)
7. C.J. Barcozy, A systematic correlation for two phase pressure drop. Chem. Eng. Prog. **62**(44), 232–249 (1966)
8. D.R.H. Beattie, S. Sugawara, Steam-water void fraction for vertical upflow in a 73.9 mm pipe. Int. J. Multiph. Flow **12**(4), 641–653 (1986)
9. D.R.H. Beattie, P.B. Whalley, A simple two-phase frictional pressure drop calculation method. Int. J. Multiphase Flow: Bried Commun. **8**(1), 83–87 (1982)
10. H.D. Beggs, An experimental study of two phase flow in inclined pipes. Ph.D. thesis, University of Tulsa, 1972
11. D. Bestion, The physical clousre laws in the CATHARE code. Nucl. Eng. Des. **124**, 229–245 (1990)
12. S.M. Bhagwat, A.J. Ghajar, Flow patterns and pipe orientation independent semi-empirical void fraction correlation for a gas-liquid two-phase flow based on the concept of drift flux model. Paper presented at the ASME 2012 Summer Heat Transfer Conference, Puerto Rico, 2012
13. S.M. Bhagwat, A.J. Ghajar, Similarities and differences in the flow patterns and void fraction in vertical upward and downward two phase flow. Exp. Thermal Fluid Sci. **39**, 213–227 (2012). doi:10.1016/j.expthermflusci.2012.01.026
14. H. Blasius, Das Anhlichkeitsgesetz bei Reibungsvorgangen in Flussikeiten Gebiete Ingenieurw 131, 1913
15. R.H. Bonnecaze, Greskovich E.J. Erskine, Holdup and Pressure Drop for Two Phase Slug Flow in Inclined Pipelines. AIChE **17**, 1109–1113 (1971)
16. C.D. Bowers, P.S. Hrnjak, Determinaiton of void fraction in seperated two phase flows using optical techiniques. Paper presented at the international refrigeration and air-conditioning conference paper 1083, Purdue, 2010
17. J. Cai, T. Chen, Q. Ye, Void fraction in bubbly and slug flow in downward air-oil two phase flow in vertical tubes. Paper presented at the international symposium on multiphase flow, Beijing, 1997
18. A. Cavallini, G. Censi, D. Del Col, L. Doretti, G.A. Longo, L. Rossetto, Condensation of halogenated refrigerants inside smooth tubes. HVAC&R Res. **8**, 429–451 (2002)
19. J.J. Chen, A further examination of void fraction in annular two-phase flow. Int. J. Heat Mass Transf. **29**, 4269–4272 (1986)
20. D. Chisholm, Pressure gradients due to the friction during the flow of evaporating two phase mixtures in smooth tubes and channels. Int. J. Heat Mass Transf. **16**, 347–358 (1973)
21. J. Choi, E. Pereyra, C. Sarica, C. Park, J.M. Kang, An efficient drift flux closure relationship to estimate liquid holdups of gas-liquid two phase flow in pipes. Energies **5**, 5294–5306 (2012)
22. R. Chokshi, Prediction of pressure drop and liquid holdup in vertical two phase flow through large diameter tubing. Ph.D. thesis, University of Tulsa, 1994

23. S.W. Churchill, Friction factor equation spans all fluid- flow regimes. Chem. Eng. **7**, 91–92 (1977)
24. A. Ciccihitti, C. Lombardi, M. Silvestri, R. Soldaini, G. Zavatarelli, Two-phase cooling experiments: pressure drop, heat transfer and burnout measurments. Energ. Nucl. **7**, 407–429 (1960)
25. A. Cioncoloni, J.R. Thome, Void fraction prediction in annular two phase flow. Int. J. Multiph. Flow **43**, 72–84 (2012)
26. N.N. Clark, R.L. Flemmer, Predicting the Holdup in two phase bubble upflow and downflow using the Zuber and Findlay drift flux model. AIChE **31**(3), 500–503 (1985)
27. C.F. Colebrook, Turbulent flow in pipes, with particular reference to the transition between the smooth and rough pipe laws. J. Inst. Civ. Eng. **11**, 1938–1939 (1939)
28. W.F. Davidson, P.H. Hardie, C.G.R. Humphreys, A.A. Markson, A.R. Mumford, T. Ravese, Studies of heat transmission thorugh boiler tubing at pressures from 500–3300 lbs. Trans. ASME **65**(6), 553–591 (1943)
29. R. De Rauz, Taux de vide et glissement dans un ecoulement biphase ascendent et descendent CENG Note TTN 165, 1976
30. A.E. Dukler, M. Wicks, R.G. Cleveland, Frictional pressure drop in two phase flow: Part A and B. AIChE **10**, 38–51 (1964)
31. K.J. Elkow, Void fraction measurement and Analysis at normal gravity and microgravity conditions. M.S. thesis, University of Saskatchewan, 1995
32. M. Espedal, K. Bendiksen, Onset of instabilites and slugging in horizontal and near horizontal gas liquid flow. Paper presented at the European two phase flow group meeting, Paris, France, 1989
33. Y.Q. Fan, Q. Wang, H.Q. Zhang, C. Sarica, Experimental study of air-water and air-oil low liquid loading horizontal two phase flow in pipes, 2007
34. X.D. Fang, Y. Xu, Z.R. Zhou, New correlations of single phase fricitor factor for turbulent pipe flow and evaluation of existing single phase friciton factor correlations. Nucl. Eng. Des. **241**, 897–902 (2011)
35. R.C. Fernandes, Experimental and therotical studies of isothermal upward gas-liquid ffows in vertical tubes. Ph.D. thesis, University of Houston, 1981
36. M. Fourar, S. Boris, Experimental study of air water two phase flow through a fracture. Int. J. Multiph. Flow **4**, 621–637 (1995)
37. H. Franca, R.T. Lahey, The use of drift flux techniques for the analysis of horizontal two phase flows. Int. J. Multiph. Flow **18**, 787–801 (1992)
38. L. Friedel, Improved friction pressure drop correlation for horizontal and vertical two phase pipe flow. Eur. Two Phase Flow Group Meet. Pap. **18**, 485–492 (1979)
39. L. Friedel, Two-phase frictional pressure drop correlation for vertical downflow. Ger. Chem. Eng. **8**(1), 32–40 (1985)
40. F. Garcia, R. Garcia, J.C. Padrino, C. Mata, J.L. Trallero, D.D. Joseph, Power law and composite power law friciton factor correlations for laminar turbulent gas liquid flow in horizontal pipe lines. Int. J. Multiph. Flow **29**(10), 1605–1624 (2003)
41. A.J. Ghajar, S.M. Bhagwat, Effect of void fraction and two-phase dynamic viscosity models on prediction of hydrostatic and frictional pressure drop in vertical upward gas-liquid two phase flow. Heat Transfer Eng. **34**(13), 1044–1059 (2013)
42. A.J. Ghajar, C.C. Tang, Advances in void fraction, flow pattern maps and non-boiling heat transfer two phase flow in pipes with various inclinations. Adv. Multiph. Flow Heat Transfer **1**, 1–52 (2009) (Chapter 1)
43. A.J. Ghajar, C.C. Tang, Void fraction and flow patterns of two phase flow in upward, downward vertical and horizontal pipes. Adv. Multiph. Flow Heat Transfer **4**, 175–201 (2012) (Chapter 7)
44. H. Goda, T. Hibiki, S. Kim, M. Ishii, J. Uhle, Drift flux model for downward two phase flow. Int. J. Heat Mass Transf. **46**, 4835–4844 (2003)

45. L.E. Gomez, O. Shoham, Z. Schmidt, R.N. Choshki, T. Northug, Unified mechanistic model for steady state two phase flow: horizontal to upward vertical flow. Soc. Petrol. Eng. **5**, 339–350 (2000)
46. D.M. Graham, T.A. Newell, J.C. Chato, Experimental investigation of void fraction during refrigerant condensation. University of Illinois, Urbana Champaign, 1997
47. E.J. Greskovich, W.T. Cooper, Correlation and prediction of gas-liquid holdups in inclined upflows. AIChE **21**, 1189–1192 (1975)
48. R. Gronnerud, Investigation of liquid holdup, flow resistance and heat transfer in circulation type of evaporators, part iv: two phase flow resistance in boiling refrigerants Annexe 1972-1, Bull de l'Inst Froid, 1979
49. A.R. Hasan, Void fraction in bubbly and slug flow in downward vertical and inclined systems. Soc. Petrol. Eng. Prod. Facil. **10**(3), 172–176 (1995)
50. A. Hernandez, Experimental research on downward two phase flow. Paper presented at the SPE annual technical conference and exhibition San Antanio, TX-U.S.A, 2002
51. G.F. Hewitt, C.J. Martin, N.S. Wilkes, Experimental and modeling studies of churn-annular flow in the region between flow reversal and the pressure drop minimum. Physicochem. Hydrodyn. **6**(1/2), 69–86 (1985)
52. T. Hibiki, M. Ishii, One-dimensional drift flux model and constitutive equations for relative motion between phases in various two-phase flow regimes. Int. J. Heat Mass Transf. **46**, 4935–4948 (2003)
53. D. Jowitt, C.A. Cooper, K.G. Pearson, The thetis 80 % blocked cluster experiments part 5: level swell experiments safety and engineering science, 1984
54. D.S. Jung, R. Radermacher, Prediction of pressure drop during horizontal annular flow boiling of pure and mixed refrigerants. Int. J. Heat Mass Transf. **32**(12), 2435–2446 (1989)
55. M. Kaji, B.J. Azzopardi, The effect of pipe diameter on the structure of gas-liquid flow in vertical pipes. Int. J. Multiph. Flow **36**(4), 303–313 (2010)
56. I. Kataoka, M. Ishii, Drift flux model for large diameter pipe and new correlation for pool void fraction. Int. J. Heat Mass Transfer **30**, 1927–1939 (1987)
57. K. Kawanishi, Y. Hirao, A. Tsuge, An experimental study on drift flux parameters for two phase flow in vertical round tubes. Nucl. Eng. Des. **120**, 447–458 (1990)
58. K.D. Kerpel, B. Ameel, C. T'Joen, H. Caniere, M. De-Paepe, Flow regime based calibration of a capacitive void fraction sensor for small diameter tubes. Int. J. Refrig. **36**(2), 390–401 (2013)
59. S.L. Kokal, J.F. Stainslav, An experimental study of two phase flow in slightly inclined pipes II: liquid holdup and pressure drop. Chem. Eng. Sci. **44**, 681–693 (1989)
60. S. Koyoma, J. Lee, R. Yonemoto, An investigation on void fraction of vapor-liquid two phase flow for smooth and microfin tubes with R134a at adiabatic condition. Int. J. Multiph. Flow **30**, 291–310 (2004)
61. C.W. Lau, Bubbly and slug flow pressure drop in inclined pipe. BS thesis, MIT, 1972
62. T.W. Lim, Flow pattern and pressure drop of pure refrigerants and their mixtures in horizontal tube. J. Mech. Sci. Technol. **19**(12), 2289–2295 (2005)
63. T.W. Lim, Y. Fujita, Flow pattern and pressure drop in flow boiling of pure refrigerants and their mixture in horizontal tube. Mem. Fac. Eng., Kyushu Univ. **62**(1), 41–54 (2002)
64. S. Lin, C.C.K. Kwok, R.Y. Li, Z.H. Chen, Z.Y. Chen, Local frictional pressure drop during vaporization of R12 through capillary tubes. Int. J. Multiph. Flow **17**, 95–102 (1991)
65. S. Lips, J.P. Meyer, Experimental study of convective condensation in an inclined smooth tube. part II: inclination effect on pressure drop and void fraction. Int. J. Heat Mass Transf. **55**, 405–412 (2012)
66. R.W. Lockhart, R.C. Martinelli, Proposed correlation of data for isothermal two phase, two component flow in pipes. Chem. Eng. Prog. **45**, 39–48 (1949)
67. A. Lorenzi, G. Stogia, Downward two phase flow: experimental investigation. Energia Nucleare **23**(7), 396–401 (1976)
68. Z.S. Mao, A.E. Dukler, The myth of churn flow. Int. J. Multiph. Flow **19**(2), 377–383 (1993)

69. R.C. Martinelli, D.B. Nelson, Prediction of pressure drop during forced circulation boiling of water. Trans. ASME **90**, 695–702 (1948)
70. L. Mattar, G.A. Gregory, Air-oil slug flow in upward inclined pipe—I slug velocity, holdup and pressure gradient. J. Can. Pet. Technol. **13**, 69–76 (1974)
71. W.H. McAdams, W.K. Woods, L.V. Heroman, Vaporization inside horizontal tubes—II benzene oil mixtures. Trans. ASME **64**, 193–200 (1942)
72. K. Minami, J.P. Brill, Liquid holdup in wet gas pipelines. Soc. Petrol. Eng. **2**, 36–44 (1987)
73. K. Mishima, T. Hibiki, Some characteristics of air-water two phase flow in small diameter vertical tubes. Int. J. Multiph. Flow **22**(4), 703–712 (1996)
74. H. Mukherjee, An experimental study of inclined two phase flow. Ph.D. thesis, The University of Tulsa, 1979
75. H. Muller-Steinhagen, K. Heck, A simple friction pressure drop correlation for two-phase flow in pipes. Chem. Eng. Process **20**, 297–308 (1986)
76. V.T. Nguyen, Two phase gas-liquid cocurrent flow : an investigation of holdup, pressure drop and flow patterns in a pipe at various inclinations. Ph.D. thesis, The University of Auckland, 1975
77. C.R. Nichols, A study of vertical flow of air-water mixture. Ph.D. thesis, University of Maryland, 1965
78. R. Oliemans, Two phase flow in gas transmission pipelines. Paper presented at the Pteroleum Division ASME Meeting, Mexico, 1976
79. N.K. Omebere-Iyari, B.J. Azzopardi, A study of flow patterns for gas-liquid flow in small diameter tubes. Chem. Eng. Res. Des. **85**(A2), 180–192 (2007)
80. O. Oshinowo, Two phase flow in a vertical tube coil. Ph.D. thesis, University of Toronto, 1971
81. M. Ottens, H.C.J. Hoefsloot, P.J. Kamersma, Correlations predicting liquid holdup and pressure gradient in steady state nearly horizontal cocurrent gas-liquid pipe flow. Inst. Chem. Eng. Trans. IChemE **79** (A), 581–592 (2001)
82. G. Paras, Characterization of downward two phase flow by neutron noise analysis. MS thesis, University of Washington, 1982
83. A. Permoli, D. Francesco, A. Prima, An emperical correlation for evaluating two phase mixture density under adiabatic conditions. Paper presented at the European two phase flow group meeting Milan, Italy, 1970
84. J.M. Quiben, J.R. Thome, Flow pattern based two phase frictional pressure drop model for horizontal tubes. Part II: new phenomenological model. Int. J. Heat Fluid Flow **28**, 1060–1072 (2007)
85. R. Revellin, J.R. Thome, Adiabatic two phase frictional pressure drop in microchannels. Exp. Thermal Fluid Sci. **31**, 673–685 (2007)
86. S.Z. Rouhani, E. Axelsson, Calculation of void volume fraction in the subcooled and quality boiling regions. Int. J. Heat Mass Transf. **13**, 383–393 (1970)
87. P.S. Sacks, Measured characteristics of adiabatic and condensing single component two phase flow of regrigerant in a 0.377 inch diameter horizontal tube. Paper presented at the ASME winter annual meeting, Houston, Texas, 1975
88. K. Sekoguchi, Y. Saito, T. Honda, JSME Reprint No. 700. 7:83, 1970
89. T.A. Shedd, Void fraction and pressure drop measurements for refrigerant R410a flows in small diameter tubes, 2010
90. D.G. Shipley, Two phase flow in large diameter pipes. Chem. Eng. Sci. **39**, 163–165 (1982)
91. S.L. Smith, Void fraction in two phase flow: a correlation based on an equal velocity head model. Inst. Mech. Eng. **184**(Part 1), 647–657 (1969)
92. A.L. Souza, M.M. Pimenta, Prediction of pressure drop during horizontal two phase flow of pure and mixed refrigerants. Paper presented at the ASME conference on cavitation and multiphase flow, HTD, 1995
93. P.L. Spedding, J.J. Chen, Holdup in two phase flow. Int. J. Multiph. Flow **10**, 307–339 (1984)

94. M. Sujumnong, Heat transfer, pressure drop and void fraction in two phase two component flow in vertical tube. Ph.D. thesis, The University of Manitoba, 1997
95. K.H. Sun, R.B. Duffey, C.M. Peng, The prediction of two phase mixture level and hydrodynamically controlled dryout under low flow conditions. Int. J. Multiph. Flow **7**, 521–543 (1981)
96. L. Sun, K. Mishima, Evaluation analysis of prediction methods for two phase flow pressure drop in mini channels. Int. J. Multiph. Flow **35**, 47–54 (2009)
97. K. Takeuchi, M.Y. Young, L.E. Hochreiter, Generalized drift flux correlation for vertical flow. Nucl. Sci. Eng. **112**, 170–180 (1992)
98. J.R.S. Thom, Prediction of pressure drop during forced circulation boiling of water. Int. J. Heat Mass Transf. **7**(7), 709–724 (1964)
99. T.N. Tran, Y.C. Chu, M.W. Wambsganss, D.M. France, Two phase pressure drop of refrigerants during flow boiling in small channels: an experimental investigation and correlation development. Int. J. Multiph. Flow **26**, 1739–1754 (2000)
100. X. Tu, P.S. Hrnjak, Pressure drop characteristics of R134a two phase flow in a horizontal rectangular microchannel. Paper presented at the international mechanical engineering congress and exposition, New Orleans, USA, 2002
101. J.M. Turner, G.B. Wallis, The separate cylinders model of two phase flow Thayer's School of Engineering, Dartmouth College, 1965
102. K. Usui, K. Sato, Vertically downward two phase flow (I) void distribution and average void fraction. J. Nucl. Sci. Technol. **26**(7), 670–680 (1989)
103. C.C. Wang, S.K. Chiang, Y.J. Chang, T.W. Chung, Two phase flow resistance of refrigerants R22, R410A and R407C in small diameter tubes. Inst. Chem. Eng., Trans. IChemE **79**, 553–560 (2001)
104. L. Wojtan, T. Ursenbacher, J.R. Thome, Measurement of dynamic void fractions in stratified types of flow. Exp. Thermal Fluid Sci. **29**, 383–392 (2005)
105. M.A. Woldesemayat, A.J. Ghajar, Comparison of void fraction correlations for different flow patterns in horizontal and upward inclined pipes. Int. J. Multiph. Flow **33**, 347–370 (2007)
106. S. Wongwises, M. Pipathttakul, Flow pattern, pressure drop and void fraction of gas-liquid two phase flow in an inclined narrow annular channel. Exp. Thermal Fluid Sci. **30**, 345–354 (2006)
107. Y. Xu, X. Fang, A new correlation of two phase frictional pressure drop for evaporating two phase flow in pipes. Int. J. Refrig. **35**, 2039–2050 (2012)
108. Y. Xu, X. Fang, X. SU, Z. Zhou, W. Chen, Evaluation of frictional pressure drop correlations for two-phase flow in pipes. Nucl. Eng. Des. **253**, 86–97 (2012)
109. D.A. Yashar, T.A. Newell, J.C. Chato, Experimental investigation of void fraction during horizontal flow in smaller diameter refrigeration applications. ACRC TR-141 University of Illinois, Urbana Champaign, 1998
110. J. Yijun, K. Rezkallah, A study on void fraction in vertical co-current upward and downward two-phase gas-liquid flow—I: experimental results. Chem. Eng. Commun. **126**, 221–243 (1993)
111. S.H. Yoon, E.S. Cho, Y.W. Hwang, M.S. Kim, K. Min, Y. Kim, Characteristics of evaporative heat transfer and pressure drop of carbon dioxide and correlation development. Int. J. Refrig. **27**, 111–119 (2004)
112. W. Zhang, T. Hibiki, K. Mishima, Correlations of two phase frictional pressure drop and void fraction in mini-channel. Heat Mass Transf. **53**, 433–465 (2010)
113. H.D. Zhao, K.C. Lee, D.H. Freeston, Geothermal two phase flow in horizontal pipes. Paper presented at the proceedings of world geothermal congress, Tokyo, Japan, 2000
114. S.M. Zivi, Estimation of steady state steam void fraction by means of the principle of minimum entropy production. Trans. ASME, J. Heat Transfer **86**, 247–252 (1964)

MIX
Papier aus verantwortungsvollen Quellen
Paper from responsible sources
FSC® C105338

If you have any concerns about our products,
you can contact us on
ProductSafety@springernature.com

In case Publisher is established outside the EU,
the EU authorized representative is:
Springer Nature Customer Service Center GmbH
Europaplatz 3, 69115 Heidelberg, Germany

Printed by Libri Plureos GmbH
in Hamburg, Germany